Anonymous

Useful Plants of Japan, Described and Illustrated.

Anonymous

Useful Plants of Japan, Described and Illustrated.

ISBN/EAN: 9783337163730

Printed in Europe, USA, Canada, Australia, Japan

Cover: Foto ©berggeist007 / pixelio.de

More available books at **www.hansebooks.com**

USEFUL PLANTS

OF

JAPAN

DESCRIBED AND ILLUSTRATED.

AGRICULTURAL SOCIETY OF JAPAN.

TAMEIKE 1, AKASAKA,

TOKYO.

1895.

USEFUL PLANTS OF JAPAN
DESCRIBED AND ILLUSTRATED.

VOLUME I.

CHAPTER I.—CEREALS & LEQUMINOUS PLANTS.

The agricultural products included under this general name consist of the most indispensable articles of human food. These grains are used as our daily food or to brew *sake* (rice-beer) or *shōyū* (soy). The straw is used for the manufacture of various articles or to feed cattle. The young pods of beans and young shoots of buck-wheet, etc. are consumed as culinary vegetables.

1. **Oryza sativa**, *L.*, aquatica. Common or Paddy rice, Jap. *Kome, Uruchi;* an annual cereal grass cultivated in paddy or marshy ground. There exist several kinds of rice, but only three kinds are usually distinguished, early, middle, and late. The grains hulled, pounded, and boiled play an important role in the Japanese alimentation as meal and porridge, or are used in the preparation of rice ferment, *sake*, and vinegar. *Dango* (dumpling), and *sembci* (a kind of cracknell) are made from the flour. The boiled rice dried makes what the Japanese call *hoshii* and is eaten grilled. Starch is also obtained from the rice, and the paste is made by boiling the flour. Besides these, rice hulls and straw serve for different uses; especially the straw is used for paper making and other manufactures.

2. **Oryza montana**, *Lour.*, Upland rice, Jap. *Okabo;* an annual cereal of the order Gramineae, cultivated in ordinary dry land. Two kinds exist, common and glutinous rice. The quality, shape, as well as use are like the paddy rice.

3. **Oryza glutinosa**, *Rumph.*, Glutinous rice, Jap. *Mochigome*; a kind of rice differing from O. sativa, *L.*, only in its colour and lack of lustre, and it is toughy and highly elastic in the nature of the meal. The grains mixed with beans of Phaceolus radiata and steamed make what the Japanese call *kowa-meshi*. *Mochi* (bread made by beating the meal in a morter), *hoshii* (dried meal), *kanzarashiko* (starch flour), cakes, and *ame* (a kind of Turkish delight) are made of this grain. The straw of this rice owing to its softness and easy manipulation is used to make ropes, mattings, straw hats, etc.

4. **Hordeum vulgare**, *L.*, Barley, Jap. *Ōmugi*; a biennial graminous plant cultivated in common dry-land. The stalk attains a height of about 3 fts. This grain is only second to rice in importance and usefulness. The grain pounded and partly crushed are used chiefly to make porridge and meal, to brew *shōyū* (soy), etc. Parched barley is used to make barley-tea. The malt of this grain is necessary to brew beer and to make *ame*. *Shinju-mugi* (literally pearl barley, the grain hulled to a pure white) is used to put in soup. The flour is baked into bread. The straw of this plant owing to its softness and high lustre is prepared to straw-mosaic, hats, and different articles.

5. **Hordeum vulgare**, *L.*, **forma mutica**, Jap. *Bōdsu-mugi*; a subspecies of Hordeum vulgare of the same quality and uses. The difference is that this one has no awn or beard.

6. **Hordeum vulgare**, *L.*, **forma nudum**, Naked Barley, Jap. *Hadaka-mugi*; a subspecies of Hordeum vulgare. Its grains are easily separated from the hulls.

7. **Triticum vulgare**, *L.*, wheat, Jap. *Komugi*; a biennial cereal, having several varieties cultivated in ordinary dry land. Its stalks grow to a height of about 3 fts. The grains are used to make *miso* (a kind of sauce in solid consistency) and to brew *shōyū* and vinegar. The wheat-meal is used to make bread, *manjū* (a small cake) & other kinds of cakes. Macaroni, vermicelli, *fu* (a kind of food), *shōfu* (starch for paste) are all made

of this grain. But the straw of this kind is rough and only used to thatch the roofs of farm houses. Besides this there are two kinds of Chinese origin, one with reddish and the other with white straw used for hat making, the latter being deemed superior.

8. Triticum vulgare, *L.,* **forma nudum,** Jap. *Bōdsu-komugi;* a subspecies of wheat having the same quality and use, but no awn to the flower.

9. Panicum miliaceus, *L.,* Common or Panicle millet, Jap. *Kibi, Uru-kibi;* an annual cereal grass cultivated in ordinary dry field. The stalks grow to a height of 5-4 fts. The white grains are used as a food in the shape of meal, porridge, or made into *dango* (dumpling).

10. Panicum miliaceus, *L.,* var. **glutinosa,** Glutinous millet, Jap. *Mochi-kibi;* a variety of the preceding. Owing to its more toughy and elastic nature it is used to make *mochi* and *dango.* It is used to brew *sake* (rice-beer).

11. Panicum italicum, *L.,* Italian millet, Jap. *Awa;* an annual cereal grass cultivated in common dry field. Its stalks attain a height of 4-5 fts. The grain is yellowish white and is used as meal and porridge *Ame* is also made of it.

12. Panicum italicum, *L.,* var., Big Italian millet, Jap. *Ō-awa;* a variety of the preceding, but only larger.

13. Panicum italicum. *L.,* var., Black Italian millet, *Kuro-awa;* a variety of Panicum italicum, *L.,* with panicles of a darker colour.

14. Panicum italicum, *L.,* var. **glutinosa,** Glutinous Italian millet, Jap. *Mochi-awa;* a variety of Panicum italicum, *L.,* with the same quality and use. Owing to its more elastic nature it is prepared to make *mochi. Sake* (Rice-beer), *shochiu* (dis-tilled spirit), and *ame* are made of this grain.

15 Sorghum vulgare, *Purse,* Guinea Corn, Jap. *Morokoshi-kibi;* a cereal grass grown in common dry land. Its stalks attain a height of 7-8 fts. The flour of this grain is used to make *mochi*, *dango* and *kanzarashi* (flour obtained by elutriating the grains during the coldest season).

16. Oplysmenus frumentaceum, *Kunth*, Crow-foot millet, Jap. *Hije ;* an annual cereal grass cultivated both in paddy and common dry field. It is the most robust kind of cereals. The stalks grow to a height of about 3-4 fts. It is consumed as meal, prorridge, macaroni, and dumpling. The grain is kept long without damage.

17. Eleusine coracana, *Gaertn.*, Finger millet, Jap. *Kamomata-kibi, Kōbō-kibi ;* an annual cereal grass cultivated in common dry field, easily distinguished by the forked shape of its panicles and its hardiness. The height of the stalk is about 1½ fts. The grain is used like Oplysmenus frumentaceum (16).

17. b. Seeds of Zizania aquatica, *L.*, Jap. *Makomo ;* the seeds of this plant mixed with rice are consumed as food by boiling.

18. Beckmannia erucæformis, *Host*, Jap. *Minogome ;* a biennial cereal grass growing wild in swamps, ponds, or marshy ground lying down on the surface of water, and forthing up its stalk to the height of about a foot. The boiled grain is eaten as food.

18. b. Panicum viride, *L.*, Jap. *Av-yagi.* In the province of *Tamba* this grass is cultivated for the sake of its grain which is used as food boiled with rice, or used to make *dango* (dumpling). It answers the same purpose as Oplysmenus frumentaceum.

19. Zea mays, *L.*, Maize, Jap. *Tomorokoshi, Koraikibi ;* an annual cereal grass cultivated in an ordinary dry land, growing to a height of about 7-8 fts. The grain is eaten either boiled as meal and porridge, or parched. Bread and cake are made of this flour. Also starch and *sake* (rice beer) are made of this grain. It is of greatest economic value. There is a variety with the name *Haje-morokoshi* to make *haje* (parched grain in a bursting state).

20. Coix lachryma, *L.* Job's tear, Jap. *Tōmugi*, *Hatomugi*; an annual cereal grass cultivated in common dry land. The stalks grow to a height of 4-5 fts. The grain pounded in a morter and cleaned is consumed as meal and *mochi*. An infusion of the parched and ground grains is used instead of tea, and is called *Kosen*.

A chinese variety of larger grains grayish brown in colour with thinner shells is more easily crushed and cleaned.

20. b Coix agrestis, *Lour.* Jap. *Judsu-dama*; a species of the preceding. The shells being much harder are used to make budhist rosaries. It is also consumed in the same way as the former. There is one another variety with the name of *Ojudsudama* which is larger and rounder.

20. c. Seeds of Bamboo, Jap. *Jinengo*; the seeds of the Bambusa senanensis and *Sudsutake* and few other kinds are used as food in the shape of flour. The seed resembles the wheat in form.

21. Glysine hispida, *Moench*, Black soy-bean, Jap. *Kuro-mame*; an annual leguminous plant cultivated in ordinary dry land. The stalk grows to a height of about 2 fts. The beans have black skin. They are eaten either boiled or parched and also used to make *miso* (a kind of sauce with solid consistency), cakes, and *natto* (a cooked beans eaten as relish to rice).

22. Glycine hispida, *Moench*, White soy bean, Jap. *Shiro-mame*; a variety of the former (21), bearing a yellowish white skin of its bean. Numerous varieties as to size, form, or duration of growth occur, and all are eaten either boiled or parched. Many important services are due to this bean. They are used to make malt, *miso* (a kind of sauce), *shōyu* (bean sauce), and *yuba* (a kind of food). The *mamenoko* (bean flour) is made of the beans and is eaten with *dango*, etc. It yields a dye called *Mame-no-go*. Oil is also pressed out from these beans. They are used in many other different ways.

23. Glycine hispida, *Moench*, var., Green bean, Jap. *Aomame;* a variety of the Glycine hispida, Moench (21) with larger seeds of greenish colour. One variety with green colour both of the skin and albumen called *Konrinzai* occurs, and is used to make *Aomame-no-ko* (green bean flour).

24. Glycine hispida, *Moench*, var., Jap. *Goishimame;* a variety of Glycine hispida, Moench, (21). This seed is flat and black, and eaten boiled.

25. Glycine hispida, *Moench*, var., Jap. *Gankuimame;* a variety of Glycine hispida, Moench, closely allied to the preceding. The beans are larger and thinner in the middle, and eaten principally boiled.

26. Dolichos cultratus, *Thunb.*, Kidney bean, Jap. *Fuji-mame, Sengoku-mame, Shakjō-mame;* an annual leguminous climber with long tendrils, cultivated in ordinary dry land. The young pods are eaten boiled. Two kinds of flowers, white and purple, exist, and the grayish white beans of the former have softer pods; those of the latter are of a purplish colour and inferior in taste, but the plant is stronger.

27. Dolichos (Lablab) vulgaris, *Smith*, var., Jap. *Ajimame, Hiramame*; a variety of the preceding with larger edible pods. The beans have white, dark purple, or other colours, and all of them are good to eat boiled.

28. Dolichos umbellatus, *Thunb*, var. **volubilis**, Jap. *Sasage;* An annual leguminous climber comprehending various varieties. The illustration represents the variety with white flowers, green pods, and white beans. The young pods are eaten boiled, and the beans are used to make white *An* (made by crushing the beans and mixed with sugar).

29. Dolichos umbellatus, *Thunb*, var., Jap. *Jiuroku-sasage;* a variety of Dolichos umbellatus with pods about 2 fts in length, which are eaten boiled when green and soft.

30. **Dolichos umbellatus**, *Thunb*, var., Jap. *Akasaskge;* a variety of Dolichos umbellatus, *Thunb*. The pods are purplish, and are used as the preceding.

31. **Dolichos umbellatus**, *Thunb*, Jap. *Hata-sasage, Kintoki-sasage;* a non-climbing variety of Dolichos umbellatus, *Thunb*, with large, flat, and oval beans. As they are of the same colour as the Phaseolus radiata are used in the same way. The leaves are consumed as a vegetable.

32. **Dolichos umbellatus**, *Thunb*, var. **seminibus albis nigris**, Jap. *Yakko-sasage*; a variety of the preceding with beans of yellowish colour and black spots. It answers the same purpose as No. 28.

33. **Dolichos bicontortus**, *Durien*, Jap. *Meganesasage;* a species of *sasage* (Dolichos) characterized by its opposite pods winding round in opposite direction in the shape of a spectacles.

34. **Phaseolus radiatus**, *L.*, var. **subtriloba**, Jap. *Bundo, Yayenari;* an annual leguminous plant cultivated in common dry land. The shape and use resemble very much the Phaseolus radiatus, but its beans are green. Its stalks grow to the height of about 1 ft. and the beans are used to make green *An* (made of crushed bean with suger) or eaten boiled mixed with rice as meal or porridge. It is also used to brew *sake* (rice beer), and to make *dango* and macaroni. It is used also to make malt called *Tōgasai* in China.

35. **Phaeolus radiatus** *L.*, Red-fruited dwarf bean, Jap. *Adsuki;* an annual leguminous plant cultivated in common dry land, consisting of numerous varieties. It grows to the height of about 2 ft. The beans are mixed with rice and eaten boiled, used to give a reddish colour to *Kowameshi* (glutinous rice boiled with the bean), made into confections, or used as washing powder instead of soap.

35. b. **Phaseolus radiatus**, *L.*, Var., Jap. *Shiro-adsuki* (*White adsuki*); a variety of the preceding with white beans specially used for making white *An* (crushed bean mixed

with sugar) and also confections or used as washing powder instead of soap.

36. Phaseolus radiatus, *L.*, var. **pendula**, Jap. *Tsuruadsuki;* a climbing subspecies of Phaseolus radiatus (35). The beans are smaller and longer than those of common Phaseolus radiatus and are used in the same way.

37. Phaseolus radiatus, *L.*, var., Jap. *Dainagon-adsuki;* a variety of Phaseolus radiatus, *L.* (35) with larger beans of clear red colour. But the use and quality are the same with the common one (35).

38. Phaseolus vulgaris, *L.*, Jap. *Ingen-mame, Gogatsu-sasage, Nido-sasage;* an annual leguminous climber cultivated in common dry land. Its young pods are eaten boiled. The beans are white, crimson, variegated, etc. All of them are eaten as vegetable.

39. Phaseolus nanus, *L.*, Jap. *Tsurunashi-ingen;* an erect standing sort of Phaseolus vulgaris, *L.* attaining the height of about 1ft.

40. Phaseolus vulgaris, *L.*, var., Jap. *Aoi-mame, Gomon-mame;* an annual leguminous climber cultivated in common dry land. The shape and colour of the bean resemble somewhat the variegated asarum leaf, hence the name of Aoi-mame or Asarum bean derived. They are eaten boiled as vegetable.

41. Pisum sativum, *L.*, Common pea, Jap. *Yendo, Yendo-mame;* a biennial leguminous climber cultivated in common dry land. The stalk grows to the height of 3-4fts. It's peas are eaten either boiled or parched or used to put in cakes.

42. Pisum sativum, *L.*, var., Jap. *Saya-yendo;* a variety of Pisum sativum with white flowers and smaller greenish peas. The pods are eaten for their softness and sweetness.

42. b. Pisum sativum, *L.*, var., Dwarf pea. Jap. *Chabo-yendo;* a variety of Pisum sativum, *L.* (43). It attains the height of about 1ft. It has the same quality as the preceding.

43. Vicia faba, *L.,* Jap. *Sora-mame;* a biennial leguminous plant growing to the height of about 2fts. The beans are eaten boiled or parched or used for making *miso* and *shōyū* (soy).

43. b. Vicia faba, *L.,* var., Jap. *Otafuku-soramame;* a kind of the preceding with flatter and larger beans and of a better taste, especially of the young ones which are soft and delicious.

44. Canavalia incurva, *D.C.,* Jap. *Tatewaki, Natamame;* an annual leguminous climber in two varieties, one with white and the other, purplish beans. The young pods of the former are preserved in salt, and the latter is eaten fresh and boiled.

45. Mucuna capitata, *Wigt.* et *Arn.,* Jap. *Osharaku-mame, Hasshō-mame;* an annual leguminous climber cultivated in common dry land. The young soft grains are eaten boiled and have a taste of Vicia faba, *L.,* but this bean contains a poisonous ingredient in a slight quantity; so it is advisable to eat moderately.

46. Arachis hypogaea, *L.,* Pea-nut, Jap. *Tōjinmame, Nankin-mame;* an annual leguminous plant cultivated in common dry land. It puts forth numerous stems in all directions under ground, and they bear nuts. They are eaten parched or used in confectionary or to extract oil.

A variety with larger nuts about 3 times bigger was introduced from America in 1873.

47. Fagopyrum esculentum, *Moench,* Buck-wheat, Jap. *Soba;* an annual cultivated plant of the order Polygonaceae found in several varieties. It grows to a height of about 2fts. The flour of buck-wheat is used for making *Soba-neri* (flour kneaded with hot water to a dough) or *Soba-kiri* (macaroni form), or made into *Kōri-soba-kiri* (Soba frozen and dried) and *Hoshi-soba-kiri* (dried soba). The grains steamed and dried are eaten boiled or made into bread or *Manju* (a small cake). Its young leaves are eaten as a vegetable, and its stalks are used to feed cattle.

Chapter II.—Leaf Vegetables.

The vegetables included under this chapter are principally those, which leaves and stems are used for culinary purposes either raw or boiled, or preserved dry or kept in brine as pickles. But there are also a great many such vegetables among chapters of Root and Flower vegetabes, Cucurbitaceous fruits, Spices and Condiments to be consumed in the same way. Those are not concerned here and will be described under their respective chapters.

48. Brassica chinensis, *L.*, var., Jap. *Mikawashimana, Tsukena;* a biennial cultivated plant of the order Cruciferœ. The length of its leaves is about 1½fts. The village of *Mikawashima*, district *Teshima*, province *Musashi*, is famous for producing the best variety, whence it derives its name. The leaves are preserved in salt as pickles from late autumn to winter. Its flower buds can also be eaten, and its panicled flowers are esteemed as cut vase-flowers among the Japanese.

49. Brassica rapa, *L.*, var. **amplexicaulis**, Jap. *Shirakukina, Hirakukina, Tōna;* this resembles much the preceding, but is shorter in height. Its yellowish white leaves have crape-like wrinkles and are eaten either boiled or as pickles preserved in salt.

50. Brassica chinensis, *L.*, var. Jap. *Komatsuna, Hatakena, Fuyuna;* a biennial cultivated plant of the order Cruciferae, with leaves growing to a length of 6-8 inches. In late winter to the spring they are used much as a culinary vegetable either boiled or preserved in salt as pickles. Its young leaves or cotyledons are used to flavour soup called *Tsumamina;* a late variety of this is called *Uguisuna*.

51. Sinapis chinenesis, *L.*, Jap. *Midsuna, Kiona;* a biennial cultivated plant of the order Cruciferœ. The petiole attains the length of about 1 ft. and comes forth in bundles of several hundreds from a root. This vegetable is used in winter and spring either boiled or as pickles preserved in salt.

52. **Sinapis chinensis,** *L.*, var., *C...ORN...ibuna* ; a subspecies of the preceding with broad leaves and without segments, and it is used for about the same purpose as others, but superior in quality. The village *Mibu* in *Kudsuno* district, *Yamashiro* province, is noted for this vegetable, whence the name is derived.

52. b. **Brassica chinensis,** *L.*, var. Jap., *Suikukina;* a biennial cultivated plant of the order Cruciferae. Its leaves resemble those of the turnip and are about 1 ft. in length. In the district of *Kamo*, province *Yamashiro*, they are extentively cultivated and preserved in salt with the name *Suikukina*.

53. **Sinapis integrifolia,** *Wild,* Jap., *Takana*, *Ō-garashi*, *Oba-garashi;* a biennial cultivated vegetable. Its leaves are full of crape-like wrinkles and of 2—3 fts. in length. In winter and spring they are much consumed as culinary vegetables either boiled or preserved in salt as pickles, and are of a very good quality.

54. **Sinapis cernua,** *Thunb,* Jap, *Karashina;* like the preceding, with leaves 7—9 inches in length, which are eaten much as pickles in winter and spring. There are black and white seeds, both used as spices and condiments or for medicinal purposes.

54. b. **Sinapis cernua,** *Thunb*, var., Jap., *Chirimenna, Irana;* a variety of the preceding with wrinkled purple leaves. The finely serrated edges look very pretty. They are eaten either boiled or as pickles preserved in salt.

55. **Tetragonia expansa,** *Ait.*, Jap., *Tsuruna;* an evergreen herb of the order Ficoideœ growing wild on the seacoast of warmer regions. It creeps over the sandy ground. It is also cultivated from seeds, and its leaves are eaten as vegetable in summer and autumn.

56. **Spinacia inermis,** *Moench*, Jap., *Hōrensō;* a cultivated biennial plant of the order Chenopodiaceæ. It is sown twice in a year, in spring and autumn. It grows to the

height of 5—8 inches. It is the vegetable of late spring and late autumn and eaten boiled. It has a very sweet flavour. The plant is diœcious, and care must be taken for collecting seeds in distinguishing the fertile seeds.

56. b. Beta vulgaris, *L.*, Jap., *Tōjisa*, *Fudansō;* a biennial cultivated plant of the order Chenopodiaceae. The seeds are sown twice in a year, in spring and autumn, and the leaves are used as vegetables in all seasons of the year, whence the name of *Fudanso* (everlasting herb). A variety with crimson tinted leaves, stems, and roots is called *Kwa-yen-sai* (Flame vegetable) or *Sangojuna* (coral vegetable) and much used for the decoration of dishes.

57. Senecio sp., Jap. *Suijenjina ;* an evergreen herb of the order Compositae. As it fears cold it is cultivated in hot beds during winter, and in spring taken out and planted in open ground. Its stalk attains a height of about 2 fts. In summer and autumn its soft and sticky leaves are consumed as vegetable.

58. Oenanthe stolonifera, *D.C.*, Jap. *Seri;* a perennial marshy plant of the order Umbelliferœ growing wild in shallow water or any damp ground, and in late winter and spring the leaves are consumed as vegatable. Those cultivated in swampy ground have petioles above 1 ft. in length.

59. Cryptotoenia canadensis, *D.C.*, Jap. *Mitsuba*, *Mitsuba-jeri;* a perennial herb of the order Umbelliferae growing wild in moist valleys, but much cultivated from seeds or by dividing the roots. In spring young leaves come forth to a height of about 1 ft. They are eaten boiled, and the roots can also be eaten fried. One variety with fine thread-like petioles and shooting in bushes to 8—10 inches high is called *Ito-mitsuba* (thread Hanewort).

60. Angelica sp., Jap. *Ashitaba*, *Hachijona ;* a triennial herb of the order Umbelliferae growing wild to a height of about 4-5 fts. In the Island of *Hachijō* it is cultivated from seeds, and the young leaves are consumed as vegetables in all seasons boiled or pickled in salt. Its large roots are eaten either boiled or made into *dango* (dumpling). There is a plant called *Hama-udo* resembling it very much in form, but very poisonous. But they can be easily distinguished, as when cut the former emits a yellow juice, while the latter does not emit any.

61. **Aralia cordata**, *Thunb*, Jap. *Udo;* a perennial plant of the order Araliaceae growing wild in mountainous districts, also much cultivated in farm lands. Its young and soft stalks are eaten as vegetable in spring and summer. The malt-*udo* is a cultivated variety about 1 ft. high. Besides this, *Kan-udo* (winter *udo*), *Nenjiū-udo* (whole year *Udo*), etc., are also cultivated.

62. **Chrysanthemum coronarium**, *L.*, Jap. *Shungiku*, *Kikuna*, *Mujinsō*; a biennial cultivated plant of the order Compositae. In autumn the seeds are sown, and the young plants are eaten in winter and spring either boiled or fresh. In summer the flower stalks shoot up to a height of about 2 fts. covered chrysanthemum-like single white flowers about 2½ inches in diameter. They are also used for floral decoration.

62. b. **Papaver somniferum**, *L.*, Jap. *Keshi;* the young plants are eaten as vegetable after being scalded.

63. **Lactuca sativa**, *L.*, Jap. *Chisa, Chishana;* a biennial cultivated plant of the order Compositae. Its seeds are sown twice in spring and autumn, and its leaves are consumed as a vegetable either boiled or raw from December to May. It grows to a height of about 3 fts.

64. **Cichorium endivia**, *L.*, Jap. *Kikujisha;* a cultivated plant resembling much Lactuca sativa, *L.* Two kinds occur, one with broad and the other with narrow leaves. In winter and spring the young leaves are eaten fresh. Its stalk grows to 2-3 fts. in height.

64. b. **Veronica anagallis**, *L.*, Jap. *Kawa-jisha;* a biennial herbaceous plant of the order Scrophulariaceae growing

wild in swampy places. In spring and summer the leaves are used instead of Lactuca sativa, *L.* It is also sown early in spring to eat its cotyledonous leaves.

65. Boltonia cantoniensis, *D.C.* (Aster cantoniensis, *Bl.*), Jap. *Yomena;* a perennial plant of the order Compositae growing wild in mountainous districts. In autumn when they have grown to about 3 fts. an umbel of blue chrysanthemum-like single flowers comes out. From late spring to summer the young leaves are eaten after passing in boiling water.

66. Cnicus nipponicus, *Max.*, Jap. *Na-azami;* a perennial herb of the order Compositae growing wild in mountainous districts attaining a height of 2—3 fts. and opening flowers in autumn. From spring to summer the young leaves are consumed as vegetable.

67. Petastes japonicus, *Miq.*, Jap. *Fuki;* a perennial plant of the order Compositae growing wild or cultivated. Its petioles grow to the length of about 2 fts. In spring and summer months they are eaten after passing in boiling water or preserved in salt. Its flower-buds owing to their flavour and agreable slight bitter taste are eaten boiled or used as condiments and spices.

67. b. Petastes japonicus, *Miq.*, var., Jap. *Akita buki;* a very large variety of the preceding, pretty enough as an ornament of its extensive round leaves, but inferior in taste as vegetable.

67. c. Senecio kœmpferi, *D.C.*, Jap. *Tsuwa-buki;* the petioles of this plant are eaten either boiled or preserved in salt as the preceding.

68. Artemisia vulgaris, *L.*, Jap. *Yomogi;* a perennial plant of the order Compositae growing wild every-where on hillsides. Its stalks grow to a height of 2—3 fts. In spring the young plants are eaten after passing in boiling hot water or used to flavour and colour *mochi* or *dango* (dumgling). Its leaves are made into *Mogusa.*

69. Amarantus mangostanus, *L.*, Jap. *Hiyu;* an annual cultivated plant of the order Amaranthaceae growing to a height of 2-3 fts. In summer and autumn months the leaves are eaten either boiled or as pickles preserved in salt. It is the vegetable of the hottest months.

70. **Salsola asparagoides**, *Miq.*, Jap. *Matsuna;* an annual plant of the order Chenopodiaceae growing wild near seacoasts and attaining a height of about 3 fts., but also cultivated in farm ground from seeds. In spring and summer months the young plants are eaten after passing in boilng water.

71. **Salsola Soda**, *L.*, Jap. *Okamiru, Okahijiki, Miruna;* an annual plant of the order Chenopodiaceae growing wild in sandy sea-shores with a long stem. It is also cultivated from seeds, and in summer and autumn months its leaves and stalks are eaten after scalding in boiling hot water.

72. **Chenopodium acuminatum**, *Wild*, Jap. *Akaza;* an annual herbaceous plant of the order Chenopodiaceae growing wild everywhere, attaining a height of about 4-5 fts. The large old stems are used as canes. Besides this, *Shiro-akaza* (white variety), *No-akaza* (field variety), and several other varieties occur of the same economic value.

73. **Brasenia peltata**, *Purch.*, Jap. *Junsai, Nunawa;* a biennial aquatic plant growing wild in old ponds and marshes. In spring and summer the young leaves covered with a mucilaginous substance are eaten fresh seasoned in vinegar.

74. **Portulaca oleracea**, *L.*, var. **sativa**, Jap. *Osuberi-hiyu;* an annual cultivated plant of the order Portulacaceae. It attains a height of about 1 ft. In spring and summer months, the leaves and stalks are eaten either raw or scalded.

75. **Rumex acetosa**, *L.*, Jap. *Sukampo;* a biennial plant of the order Polygonaceae growing wild in mountainous regoins and attaining a height of about 2 fts. In spring the young soft leaves and stalks are eaten either boiled or preserved is salt. They have a pleasant aciduous flavour.

75. b. **Polygonum cuspidatum**, *S.* et *Z.*, Jap. *Itadori;*

the young stalks of this plant (905) growing wild in mountainous districts are eaten either boiled or raw, or preserved in salt, when they have attained a height of about 1 ft. and 1 inch in diameter.

76. Acanthopanax spinosum, *Miq.*, Jap. *Ukogi;* a deciduous shrub of the order Araliaceæ attaning a height of 7-8 fts. Its thorny branches fit well for hedges. In spring the young leaves are eaten scalded.

77. Aralia chinensis, *L.*, Jap. *Tara-no-ki;* a deciduous shrub of the order Araliaceae growing wild on mountainous regions and attaining a height of about 10 fts. covered with sharp thorns. In spring young leaves are eaten after passing in boiling water. This is the best of edible tree shoots.

78. Clethra barbinervis, *S.* et *Z.*, Jap. *Riyobu*, *Hatatsumori;* a deciduous tree of the order Ericaceae growing wild in mountainous districts. Its stem attains a height of about 10 fts. In spring the young leaves are eaten boiled with rice. It is much used by peasants of remote mountain villages.

79. Helwingia rusciflora, *Wild*, Jap. *Hana-ikada;* a deciduous shrub of the order Garryaceae growing wild in mountainous districts to a height of about 8 fts. In spring young soft leaves are eaten boiled.

80. Allium fistulosum, *L.*, Onion, Jap. *Negi;* a perennialherbaceous plant of the order Liliaceae. The tubular leaves grow about 2 fts. in height, and the length of white underground part varies according to the skill of cultivators. They are eaten either boiled or fresh in all seasons, but they are best and sweetest in winter.

80. b. Allium fistulosum, *L.*, var., Winter onion, Jap. *O-negi*, *Ippon-negi;* a variety of the preceding, but larger and standing erect solitally and is wholesome and sweet.

81. Allium fistulosum, *L.*, var., Jap. *Iwa-tsuki-negi;* a variety of the preceding smaller in size. Owing to its good taste it is much valued as a special product of the district *Iwatsuki* in the province of *Musashi*.

82. **Allium esculentum**, *L.*, Jap. *Wakegi, Fuyunegi;* a kind of Allium with long slender leaves. The name of *Wakegi* is derived from its benig easily propagated by division. In spring and winter months when they shoot out are eaten for their good taste and less odour.

83. **Allium** sp., Jap. *Karigi, Natsu-negi;* a smaller species of Allium fistulosum, *L.* (50). As it shoots out in summer it is used as a vegetable in that season.

84. **Allium ledebourianum**, *Schult*, Jap. *Asatsuki;* a species of Allium resembling Allium fistulosum, *L.*, growing wild, but also much cultivated. The leaves are long and slender, with small bulbs. In spring the leaves shoot out luxuriously to a length of about 1 ft. Both leaves and onions are eaten together, and the taste resembles that of Allium fistulosum, *L.*, being less odourous, but much soft and smooth.

85. **Allium nipponicum**, *Fr.* et *Sav.*, Jap. *Nobiru;* a small kind of Allium growing wild with leaves about 1 ft. long, forming small onions at foot. In spring and summer both leaves and onions are eaten together.

86. **Allium odorum**, *L.*, Jap. *Nira;* a leek much resembling allium fistulosum, *L.* (80). In spring the flat leaves come out luxuriantly from old onions, attaining a length of about 1 ft. In summer and autumn they serve as a vegetable.

87. **Colocasia antiquorum**, *Schott*, Jap. *Tō-no-imo, Aka-imo;* 'a cultivated tuberous plant of the order Araceae. Its petioles grow to a length of about 5 fts. with expanded leaves at the top, and are eaten boiled and also preserved dried or in salt. This variety does not produce many young tubers, but the mother tuber grows to a considerable size. It is an excellent article of food for its sweet taste.

87. b. **Colocasia antiquorum**, *Schott*, var., Jap. *Midsu-imo;* a close ally of the preceding cultivated in swampy

ground in the warm regions. The petioles are principally used for food.

88. Colocasia indica, *Kth.* (Caladium esculentum, *Sieb.*), Jap. *Hasu-imo;* a species very nearly related to the preceding, with leaves growing to a length of about 3-4 fts. It is called *Hasu-imo* (Lotus caladium), because the leaf has the form of Lotus-leaf. As the petioles of this kind are not bitter they are eaten either raw or boiled and also preserved dry.

89. Smilax herbacea, *L.*, var. **nipponicum,** *Miq.*, Jap. *Shiode;* a perennial wild climber of the order Smilaceae. The young leaves are boiled and eaten as a vegetable.

90. Osmunda regalis, *L.*, var. **japonica** Jap. *Jenmai;* a perennial herbaceous plant of the order Filices growing wild on mountains and woody places and forming a large clump. In spring young coiling fronds are eaten boiled or preserved dry or in salt. The white fibres covering the young fronds are woven into clothes.

91. Pteris aquilina, *L.*, Jap. *Warabi;* a perennial herbaceous plant of the order Filices. Its rhizomes extend in all directions under ground and shoot up young leaves everywhere, and are eaten boiled or preserved in brine. Starch is also got from the rhizomes, called *Warabiko* or brakefern meal. The fibre of the rhizomes after the meal emptied out are used to make rope of a dark brownish colour, bearing against rottening by moist.

91. b. Botrychium ternatum, *Sw.*, Jap. *Hana-warabi Fuyu-warabi;* a perennial plant of the order Filices growing wild on mountainous districts. In autumn the leaves come out with flower stalks, one to each leaf generally. They attain a height of 6-7 inches and are eaten boiled with a soft delicious taste.

92. Ceratopteris thalictroides, *Brong,* Jap. *Midsu-warabi, Midsu-ninjin;* an annual herbaceous plant of the order Filices growing wild in moist swampy plases. Those growing in water get larger than those grown in dry land. In late spring the young leaves are boiled and eaten as vegetable.

Besides those mentined in this chapter there are innumerable other plants with edible leaves. For instance, in the chapter of Cereals and Leguminous plants, the cotyledons of Glycine hispida, *Moench* (21-23) and Phaseolus radiata, *L.* subtriloba (34), and the young leaves of Dolichos umbellatus, *Thunb*, var. volubilis (28-33), Pisum sativum, *L.* (41-42), Vicia faba, *L.* (42), Fagopirum esculentum, *Mœnch*, etc. are eaten as vegetables. Among the chapter of Root vegetables, the young leaves of almost all plants except 2-3 are used for the same purpose, as Raphanus sativus, *L.* (93-102), Brassica rapa, *L.* (103-106), Daucas carota, *L.* (107), Lappa major, *Gaertn.* (108), Batatus edulis, *Chois,* and Solanum tuberosa, *L.* Among the Flower vegetables, the leaves of Pyrethrum sinense, *Sabin*, and the young shoots of Amomum Mioga, *Th.* or Zingiber mioga, *Roscoe;* among Cucurbitaceous plants the petioles of Cucurbita pepo, *L.* (129-130); the young leaves and stalks of several plants contained in the chapter of Spices and Condiments, as Eutrema wasabi, *Max.* (162), Raphauns sativus, *L.* (167), Perilla arguta, *Benth* (170), Capsicum longum, *L.* (167), and Polygonum nodosum, *L.*; among Starch yielding plants the young leaves of Erythronium deniscanis, *L.* (257) and Orichia edulis, *Miq.* (258); among Economic plants the soft young leaves and stalks of Kochia scoparia, *Schrad* (300) and Luffa petola, *Scr.* (305); among Oil and Wax plants the leaves of Brassica chinensis, *L.* (308); among Textile plants the young leaves of Typha japonica (339) and Zizania aquatica, *Miq.* (346); among Dye plants the leaves of Basella rubra, *L.* (371); among Medicinal plants the young leaves of Malva pulchella, *Berttn.* (406) and Plantago asiatica, *L.* (448) and the petioles of Rheum undulatum, *L.* (453); among Timber trees and Bamboos the young leaves of Cedrela chinensis, *Juss*; among Evergreen trees the young leaves of Cycas revoluta, *Thunb*; among Ornamental garden trees the young soft leaves of Althœa rosea, *Cav.* (769), Scabiosa japonica, *Miq.* (809), Taraxacum officinale, *Wigg.* (853), Adenophora verticellata, *Fisch*(861), Conandron ramondioides, *S.* et *Z.* (959) and Funkia sieboldiana, *Hook* (960) are all eaten as vegetables. Besides these, among Trees and Shrubs there are many serving to the alimentation in unexplored cold and hot climates, but these are mostly mere substitutes of food in time of famine or for quriosity, and are therefore not mentioned in this section.

CHAPTER III.—ROOT VEGETABLES.

This Chapter includes all vegetables which roots or bulbs serve for alimentation and are principally consumed in fresh state in spring and summer or preserved dried or salted for the use of other seasons. Some containing large quantities of starch are eaten in place of cereals. The leaves and stalks of some are also used as vegetables.

93. Raphanus sativus, *L.*, Jap. *Daikon, Sudsushiro;* a biennial cultivated plant of the order Cruciferae with many varieties. The species painted in the illustration is the *Nerima* raddish grown in the *Musashi* province. Of this species two forms exist, one called *Naga* (long), and the other *Tomari* (stopped). The former is about two feet long and two inches in diameter, tapering toward the end and without a tap-root. The other is about the same length, club-shaped, and has a long tap-root. It is eaten boiled or preserved in rice-bran and brine. Both roots and leaves are used as food fresh or dried.

94. Raphanus sativus, *L.*, var., Jap. *Sakurajima-daikon;* this is the largest kind of raddish. It is a speciality of the place called *Sakurajima*, province *Osumi*, whence it derives the name. There are three varieties, early, middle, and late, the last of which is the largest. It is about 3 fts. in circumference and weighs 20–30 lbs. It is thick in middle and tapers slightly towards both ends. It is eaten raw, boiled, dried, or preserved in salt, and has a sweet wholesome tast.

95. Raphanus satiavus, *L.*, var., Jap. *Azami-daikon, Suikwa-daikon;* a variety of raddish with numerous segments on leaves. The roots and leaves are used like those of the former.

96. Raphanus sativus, *L.*, var., Jap. *Otafuku-daikon*, *Kameido-daikon;* a variety of raddish (93). The special place called *Kameido* in the district *Minamikatsushika*, *Musashi* province, produces the best quality, whence the name is derived. It consists of three varieties, early, middle, and late. The middle variety which is taken up about the month of May has a root about 1⅓fts. long and 1½ inches in diameters and it is greenish at the top where the leaves come out. It is superior in taste, and eaten raw, boiled, or preserved in salt.

96. b. Raphanus sativus, *L.*, var., Jap. *Hosone-daikon;* a slender kind of Raphanus sativus, *L.*, (93), being a foot in length with a diameter of about ½ inch. This is well sown at any time, affording fresh vegetable at any time in the year, whence it is called *Toki-shiradsu* (non aware of time). It is also called *Otafuku*, but is quite different from that grown in *Kameido* of the same name (96). It is good to eat raw, boiled, or preserved in salt.

97. Raphanus sativus, *L.*, var., Jap. *Ji-daikon* or *Tokuri-daikon;* a variety of raddish (93), being club-shaped thick at lower end and about a foot long. It is good to eat boiled.

98. Raphanus sativus, *L.*, var., Jap. *Miyashige-daikon*, *Owari-daikon;* a variety of raddish (93). This is specially produced in *Miyashige*, district *Nishikasugai*, province *Owari*. It is thick at the top tapering towards the tip without a tap-root, and about 1½fts. in length and 3½ inches in diameter. It is the sweetest of raddish and the best to be boiled, preserved dry, or pickled. It soon loses its fine quality when cultivated in other districts.

98. b. Raphanus sativus, *L.*, var., Jap. *Hōrio-daikon;* a variety of raddish (93), the special product of the village *Horio*, district *Kaito*, province *Owari*. There are two varieties, one greenish at the head, and the other all white. The latter is better in quality. It resembles much *Miyashige-daikon* in shape (98 a), but with a tail-like root. It is of a very large size with a length of 2fts. and a diameter of 8 inches, and rivals the early variety of *Sakurajima-daikon* (94). It is consumed as the preceding.

98. c. Raphanus sativus, *L.*, var., Jap. *Shogoin-daikon;* a close ally of the preceding with a larger diameter. It is about a foot long and 6-7½ inches round, and of the superior flavour and taste. This is a variety got from the seed of 98. b sown in the village of *Shogoin* in the district of *Atago*, province *Yamashiro*.

99. Raphanus sativus, *L.*, var., Jap. *Hadano-daikon, Moriguchi-daikon;* a slender shaped variety of the raddish (93). There are different varieties with a length of 3-4 fts. and a diameter of about a inch. The districts of *Hadano* in the province *Sagami* and of *Moriguchi* in the province *Kawachi* are famed for this product. It is hard and better in taste. This raddish pickled in *sake*-lee is called *Moriguchi-dsuke* (*Moriguchi*-pickle). Dried it is called *Mino-boshi* (dried raddish of the province *Mino*). It is eaten boiled or preserved in a mixture of vinegar and soy.

100. Raphanus sativus, *L.*, var., Jap. *Sangatsu-daikon, Ninengo;* a variety of the raddish (93). The seeds are sown at the end of autumn, and the roots are eaten at the end of spring, being white, thin, and hard.

101. Raphanus sativus, *L.*, var., Jap. *Natsu-daikon;* a variety of *Sangatsu-daikon* 100). It is sown in spring and eaten in summer.

102. Raphanus sativus, *L.*, var., Jap. *Aka-daikon, Murasaki-daikon;* a variety of raddish (93). The leaves and roots are purple tinted. There are summer and autumn varietes originated from the common raddish.

102. b. Raphanus sativus, *L.*, var., Jap. *Sangoju-daikon;* a turnip-shaped variety of raddish (93). The outside is smooth and light crimson; the flesh is white. It has a diameter of about 2½ inches. It is planted in a pot as an ornamental plant for the new year. It is eaten raw or preserved in salt.

103. Brassica rapa, *L.*, Jap. *Kokabura ;* a biennial cultivated plant growing in ordinary dry ground, belonging to the order Cruciferae, with numerous varieties of different colours and shapes. The variety painted in this volume is one commonly cultivated in the eastern provinces. The root has a diameter of 1—1½ inches. In winter months the roots as well as leaves are eaten boiled or preserved as pickles.

104. Brassica rapa, *L.*, var., Jap. *Ōmi-kabura, Suwari-kabu ;* a variety of Brassica rapa, *L.*, of an enormous size. It is round and flattish with a diameter of about a foot. The province of *Omi* is praised for its product, whence derived its name. It is sweet and wholesome and good to eat boiled, pickled in salt or in *sake*-lee, or dried. The leaves and stalks are also preserved dry and used as vegetables. The variety got from this seed in the village of *Shogoin* in the district of *Atago*, province *Yamashiro*, is named *Shogoin-kabura* and praised for its good taste.

105. Brassica rapa, *L.*, var., Jap. *Tennoji-kabura ;* a variety of Brassica rapa, *L.*, the special producet of the village of *Tennoji, Settsu.* The root is round and somewhat flattened. It has a diameter of about 8 inches. It is soft and brittle in quality. It is cut into thin slices and pickled in salt or in vinegar seasoned with soy, or eaten raw, and especially good to preserve dried.

105. b. Brassica rapa, *L.*, var., Jap. *Murasaki-kabura, Hino-kabura ;* a variety of Brassica rapa, *L.* (103) like *Tennoji-kabura* (105) with purpish leaves and stems. The rind of the root is of a deep purple ; the flesh is white. It is pickled in salt. The village of *Hino* in the district of *Higami, Omi,* is praised for this produce, whence derived its name.

105. c. Brassica rapa, *L.*, var., Jap. *Aka-kabura, Beni-kabura, Hino-kabura ;* a variety of Brassica rapa, *L.* (103). The root is round and flat with a diameter of about 4 inches. The rind as well as the interior is of a bright crimson colour ; the leaf-stalks are shaded with crimson. When pickled in salt it gets a more bright red colour, whence called *Hino-kabura* (crimson turnip). The village *Saiin-Takehara* in the district of *Onsen*, *Iyo*, is famed for a fine strain of the variety.

106. Brassica rapa, *L.*, var., Jap. *Naga-kabura ;* a variety of Brassica rapa, *L.*, (103) with a long raddish-like shape and of a length of about a foot, thicker towards the end. It is soft and wholsome, and best to eat boiled or preserved as pickles in salt.

106. b. Brassica rapa, *L.*, var., Jap. *Momiji-kabura ;* a variety of the Brassica rapa, *L.* The root is about 9 inches long with a diameter of 1-1¼ inches. The rind is bright pink; the flesh is white. It is much cultivated in the neighbourhood of *Hikone, Omi*. When it is dried or pickled, the flesh turns crimson. It is also used to decorate dishes.

107. Daucus carota. *L.*, Jap. *Ninjin, Ninjinna ;* a biennial cultivated vegetable of the order Umbelliferae. The root, about 1½ fts. long and 1¼ inches in diameter, is orange yellow. In winter and spring months it is eaten boiled or used raw cut into slices with other food as a *Namasu*, etc. It is also preserved in *miso* (a kind of soy of a solid consistency) or in *sake*-lee. In autumn months the young soft leaves are eaten boiled.

One variety called *Kintoki-ninjin* has a bright crimson colour and is of a larger size. Its taste is sweet and wholesome. It is much grown in the neighbourhood of *Osaka*. Another variety called *Murasaki-ninjin* (purple carrot) is of a deep purple colour outside and yellow in the center.

108. Lappa major,[1] *Gaertn.*, Jap. *Gobō ;* a biennial cultivated vegetable of the order Compositae. The seeds are sown twice in a year, in spring and autumn. The root has a length of about 2½ fts. and a diameter of 1-1¼ inches. A very large kind called *Mumeda-gobo* is cultivated at *Mumeda* in the district of *Saitama, Musashi*. This is eaten boiled or preserved in salt. The young soft leaves are also eaten boiled.

109. Batatus edulis, *Chois,* Jap. *Satsuma-imo, Kara-imo, Riukiu-imo;* a cultivated tuberous creeping plant of the order Convolvulaceae and of several varieties according to the shape and colour of the tubers. They are eaten raw, boiled, baked, or steamed, and are the most important food next to cereals. They are also cooked to make cakes. They are cut into slices and preserved dry, and also reduced to flour to make *dango* (dumpling). Starch, *ame* (a kind of Turkish delight), and *sake* (ric beer) are also made. The young leaves are eaten boiled as vegetables.

109. b. Solanum tuberosum, *L.,* Jap. *Jagatara-imo;* a cultivated tuberous plant of the order Solanaceae. There are two sorts, white and red skinned. They are eaten boiled or steamed and are preserved dry. *Miso, shōyu,* and *sake* are prepared from them. They yield also starch. The young leaves can be eaten, while the young shoots are very poisonous.

110. Dioscorea Batatus, *Dcne.* Jap. *Tsukune-imo;* a cultivated tuberous creeping plant of the order Dioscoreaceae. The underground tuber is large, solid, irrerular, flat, about 9–10 inches in diameter, and elastic. As it contains a large amount of starch and is wholesome in taste, it is eaten simply boiled or steamed or as *Tororo* (a kind of gruel made by grinding the fresh tuber). It is also used in various other ways in cooking and confectionary. It is dried and made into meal. Starch is obtained from it. *Icho-imo* and *Ise-imo* are its varieties.

111. Dioscorea japonica, *Th.,* Jap. *Jinenjo, Yamano-imo;* this is the typical species of the former (110) growing wild in hills and mountains. It produces cylindrical tubers 5–6 fts. long. It is superior in quality, but used quite differently.

112. Dicscorea japonica, *Th.,* var., Jap. *Naga-imo;* the cultivated form of the preceding. The tubers have the same shape, but are shorter. They are 3–4 fts. long.

113. Dioscorea japonica, *Th.,* var., Jap. *Ichinen-imo, Lakuda-imo;* its tubers ripen in one year, attaining only to 1–2 fts. They are watery and inferior in quality.

113. b. Tubers of Dioscorea japonica, *Th.*, Jap. *Muka-go, Nukago ;* the small tubers about the size of the thumb grown at the leaf-exils of Dioscorea japonica, *Th.* (110-111), etc. are eaten boiled or planted as the seeds.

114. Colocasia antiquorum, *Schott.*, Jap. *Sato-imo, Hatake-imo ;* a cultivated tuberous plant of the order Araceæ embracing several varieties. It resembles very much *To-no-imo,* but is quite green instead of purple. The petioles grown to the length of 3-4 fts. are eaten boiled or preserved dried and eaten as vegetable.

115. Colocasia antiquorum, *Schott.*, var., Jap. *Yatsu-gashira ;* a variety of *To-no-imo* (87). The leaves come out in bundles of 8 or 9 inches with thin long petioles from the single tuber. The latter grows to the size of 5-6 inches in diameter, consisting of several sprouts, but very rarely produces young tubers. It is eaten simply boiled or steamed. The taste is wholesome resembling that of *To-no-imo.*

116. Colocasia antiquorum, *Schott.*, var., Jap. *Yegu-imo, Hana-imo ;* a variety of *Sato-imo* (114). Because of its strong acridity it is grown under thick heaps of dust in malt state, which is eaten by the name of *Ne-imo* (yam-root). The young tubers are likewise eaten boiled.

117. Conophallus konjak, *Schott.*, Jap. *Konniyaku-imo ;* a cultivated tuberous plant of the order Araceae. The tuber forms a round ball extremely acrid in taste in fresh state. The people used to make *Konniyaku*, a gelatinous tough food, by passing the raw tubers in boiling hot water, but they now make it by reducing the dried tubers into flour.

118. Sagittaria sagittifolia, *L.*, Arrow-head, Jap. *Kuwai, Shiro-kuwai ;* a bulbous plant cultivated in swampy soil and belonging to the order Alismaceæ. Several stalks sprout from one root and produce one tuber to each at the bottom, and in winter months the tubers are collected and eaten thoroughly boiled. They are also used for several purposes in cooking. Starch is obtained from the bulbs. Generally the tuber is of a diameter of 1¼ inches and sometimes larger. Besides this there is a Chinese kind with long oval tubers.

119. **Sagittaria sagittifolia**, *L.*, Jap. *Suita-kuwai, Omodaka, Gowai;* a small kind of *Kuwai* (118) growing wild in swampy field, but often cultivated for the *sake* of its tubers. It is about half a inch in diameter and is eaten boiled.

120. **Scirpus tuberosus**, *Sim.*, Jap. *Kuro-kuwai;* a bulbous plant of the order Cyperaceae growing wild in marshy places. It is also cultivated in paddy land for the sake of its tubers. In winter they are dug out and eaten either raw or boiled. They resemble the chestnut in taste. In China starch is made from them and called *Batei-fun*.

121. **Lilium tigrinum**, *Gawl.*, Tiger lily, Jap. *Oni-yuri; Ryori-yuri;* a cultivated bulbous plant of the order Liliaceæ. In winter the bulbs are taken up and eaten boiled and cooked. It is very sweet and wholesome, the flavour varying with the soil. This variety produces in the axils of leaves small bulbs with which we can propagate the plant. The wild growing variety is also eaten. The stem grows to a height of 3-4 fts., blooming many flowers which are very beautiful.

121. b. **Lilium auratum**, *Lindl.*, Jap. *Yama-yuri, Sasa-yuri;* this bulbous plant (942) is much praised for the beauty of its flowers and also much cultivated for the sake of its bulbs. In winter when the bulbs have grown to a great size, they are taken up and consumed as vegetables. The flowers are much valued for their fragrance and the beauty of their colours.

121. c. **Lilium elegans**, *Thunb.*, **Lilium thunbergianum**, *Roem.* et *Schult.*, Jap. *Hime-yuri, Hi-yuri, Sukashi-yuri;* this kind is much cultivated for its edible bulbs for summer. The bulbs are about 1½ inches in diameter and pure white. It contains no bitter principle. The ornamental flower-lity, Lilium concolor (949), has the same Japanese name, but is quite different.

122. **Allium sativum**, *L.*, Jap. *Ninniku, Hiru;* a bulbous cultivated plant of the order Liliaceæ, of the same family as

Allium fistulosum. It forms the bulb of a strong pungent odour. In spring both the leaves and bulbs are eaten.

123. Allium splendens, *Wild.* (A. arenarium, *Thunb.*, A. bakeri), Jap. *Rakkio, Giyoja-biru ;* it belongs to the same family as the preceding (122). The small bulbs of the size of a thumb are eaten boiled or preserved as pickles in an air tight vessel in a boiled mixture of *sake* (rice-beer), vinegar, and soy They are eaten after two months thus steeped.

124. Stachys sieboldi, *Miq.*, Jap. *Cho-rogi, Chiyo-rogi ;* a cultivated biennial plant belonging to the order Labiatae Many white rosary-like tubers of the size of a finger grow attached to the root. In winter these tubers are taken up and eaten boiled or preserved in salt or *Ume-dsu* (plum-vinegar). Those preserved in the latter juice is very good and beautiful being pink coloured.

124. b. Lycopus lucidus, *Turcz.*. Jap. *Shirone, Ajekoshi ;* a perennial plant of the order Labiatae growing wild in swamps or near ponds. Its white rhizomes grow to a length about 1ft., and thicken towards the end where it reaches to the size of a finger. It is knitted and of the same shape as the preceding, but of a length of 5–6 inches. In winter they are eaten boiled or preserved in salt.

125. Rhizome of Nelumbo mucifera., Jap. *Hasu-no-ne, Renkon ;* a perennial plant of the order Nymphaeaceae cultivated in swamps and marshes. The rhizomes lie far down in the mud and grow to a length of 3–4 fts. They are cylindrical, white, and consist of a succession of joints. The interior is permeated with about 10 canals. In winter and spring they are dug out and eaten boiled or presewed in sugar or reduced into starch called *Hasu-no-ko* (Lotus-meal). One kind recently introduced from China has thicker irregular rhizomes with small holes at the joints, and is wholesome in taste. The young leaves are likewise eaten as vegetables. The fruits and flowers are respectively described in the devision of fruits (228) and ornamental plants (149).

125. b. Bamboo sprout, Jap. *Take-no-ko, Takanna ;* the

bamboo sprouts are the young soft stems of bamboos belonging to the order Gramineae. Almost all kinds of bamboos are edible, but *Moso-chiku* is noted for its good taste. Next to this are *Hachiku*, *Hotcichiku*, *Madake*, etc. These are eaten boiled, or preserved in salt or dried. The young sprouts are clothed with a sheath which is taken off after the sprout has grown to full size and is used for various purposes. Those of *Madake* and *Hachiku* are most extensively used.

125. c. Small bamboo sprouts, Jap. *Haimo, Yokotake, Muchiko;* a young bamboo sprout growing up obliquely from the end of bamboo roots, resembling the preceding one in shape and taste. They grow at all seasons of the year, but in autumn they are mostly produced. For the sake of propagating the bamboo wood, it is advisable not to take them out of the ground.

125. d. Young shoots of Phragmites roxburghii, *Kunth*, Jap. *Yoshigo, Yoshi-dsuno;* the young shoot of this plant resembles the small sized bamboo sprout and is eaten in the same way. In China this shoot taken out of the sheath is dried and preserved with a coating of salt on it, and used for various cooking purposes.

125. e. Asparagus lucidus, *Lindl.*, Jap. *Tenmondō;* tubers growing together about the size of a finger are preserved in sugar, or used for various cooking purposes after having been boiled in water to take away the acidity.

Note.—The plants mentioned in this chapter are some of the principal kinds which roots are used as vegetables, but there are many other different varieties produced in different places; for example in the sweet potato there are many varieties produced in warm countries; such are too numerous to be mentioned in this limited space. The bamboo sprout though really not a root vegetable is temporally put in this division.

CHAPTER IV.—Flower Vegetables.

This chapter includes plants which petals, buds, and peduncles are eaten as vegetables. They are of a limited number. Most of them are cooked after scalding in boiling water. They are eaten rather as a curiousity of culinary herbs; some of them are used partly for condiments and spices.

126. Pyrethrum sinense, *Sabin,* Jap. *Riyori-giku;* a perennial cultivated plant belonging to the order Compositae. There are two kinds; one blooms only in autumn, and the other in summer and autumn. The former attains a height of about 2fts. and the latter, a little shorter. Their yellow flowers are eaten cooked after slightly boiled in water. They are also dried and preserved. Their leaves are likewise eaten when fried.

127. Equisetum arvense, *L.,* Jap. *Tsukushi;* a perennial plant of the order Equisetaceae, growing wild in fields. In the beginning of spring before the flower-stalk produces its spores, it is eaten boiled, preserved in salt, or put in vinegar mixed with soy after havig been boiled in water.

128. Amomum mioga, *Th.,* Jap. *Myōga;* a perennial plant of the order Zingiberaceae growing wild. It is also cultivated. It grows to a height of about 3fts. There are two with red kinds and white roots. In summer and autumn the flowers with the bracts are taken and eaten either raw or boiled. It has an aromatic odour with a slight acid taste. The old leaves when twisted and kneaded are used for making saddles.

128. b. Brassica flowers, Jap. *Na-no-hana;* all the flowers of the Brassica family are edible as vegetables, especially flowerbuds of Brassica chinensis.

128. c. Flowers of Petasites japonicus, *Miq.,* Jap. *Fuki-no-tō;* the flower-buds of Petasites japonicus, *Miq.,* are eaten either raw or boiled on account of their aromatic bitterness.

128. d. Flowers of Paeonia moutan, *Sims.,* Jap. *Botan-no-hane;* the petals of white and pink peony flowers are eaten cooked after boiling, and those of Paeonia officinaris likewise.

— 31 —

128. e. Flowers of Gardenia florida, *L.*, Jap. *Kuchinashi-no-hana;* the six parted monopetalous fragrant flowers are eaten after having been boiled and cooked.

128. f. Flowers of Hemerocallis flava, *L.*, Jap. *Kuwanso-no-hana;* the flowers of Hemerocallis flava, *L.* (955) are eaten when slightly boiled or preserved. The young leaves are likewise eaten as a vegetable. Almost all the flowers belonging to this species are edible ; the buds of thin single flowering sorts are much used for cooking purposes in China, and are called *Kinshinsai* which are prepared by drying after having been slightly boiled.

128. g. Flowers of Prunus pseudo-cerasus, *Lindl.*, Jap. *Sakura-no-hana;* the double flowering cherries, especially those which petals do not readily fall off, are preserved in salt and prepared to a drink like tea. The variety called *Yedozakura* is often selected for this purpose.

128. h. Flowers of Cymbidium virens, *Lindl.*, Jap. *Ran-no-hana;* the flowers of some orchids are often preserved in salt and put in hot water, being used as a drink. Especially the flowers of *Hokuri* (914), Cymbidium virens, Lindl., is used for this purpose. It is also preserved in plum-vinegar.

CHAPTER V.—Cucurbitaceous Fruits.

This chapter includes herbaceous plants which fruits are consumed as vegetables principally in summer and autumn, such as melons, cucumbers, egg-plants, etc. They are eaten either raw, boiled, baked, or preserved in salt or sugar according to their nature. Besides those mentioned in this chapter, there are some other cucurbitaceous fruits as well as fruits produced from trees used as vegetables. They are mentioned in the chapter of fruits.

129. Cucurbita pepo, *Linn.*, Jap. *Tōnasu, Bōbura, Nankin;* an annual cultivated climbing plant belonging to the order Cucurbitaceae. It consists of different sizes of fruits which are eaten boiled or dried and preserved. In autumn and winter their petioles are likewise eaten as vegetables. The one here mentioned is a flat variety called *Naitō-tōnasu*.

130. Cucurbita pepo, *Linn.*, Jap. *Kabocha;* this is a variety of the preceding (129). Its shape resembles that of the gourd and is used in the same way.

131. Cucurbita aurantia, *Linn.*, Jap. *Kintōga, Benitōga, Akatōgau;* it is of the same family as the preceding of a long oval shape and red lustrous skin. It is rather insipid, and so cannot be eaten ; it is only used as an ornament in fruit stores on account of its beautiful appearance.

132. Cucumis common, *Th.*, Jap. *Shiro-uri, Asa-uri;* an annual cultivated climbing plant of the order Cucurbitaceae. The fruit is light green or almost entirely white with an oblong oval shape about a foot long. It is eaten raw or boiled, preserved in salt or *sake*-lee, or dried. There are also black and green varieties.

133. Lagenaria dasistemon, *Miq.* Jap. *Tōga, Tōgan, Kamouri;* an annual climbing plant cultivated in farms belonging to the order Cucurbitaceae. The melon is oval and a ft. in diameter. The skin is covered with fine hair and white powder. It is preserved fresh or in sugar for the use in autumn and winter months. The young soft melons when they have grown to about 2-3 inches in diameter are used for various cooking in *Osaka* and are called *Chōsen-uri* (Corean melon). Another variety about 3 fts. in length grows in *Kiushiu*. It is smooth and lustrous without white powder.

134. Cucumis sativus, *L.*, var., Cucumber, Jap. *Kiuri;* an annual climbing cucurbitaceous plant cultiatved in fields or forced in hot beds to have melons in early spring. The fruits are oblong oval, and provided with small numerous protuberances. They are eaten raw or roasted when they are about 2-7 inches long. They are also preserved in salt or in bran. The fruits when ripe are eaten boiled.

134. b. Cucumis sativus, *L.*, var., Jap. *Naga-kiuri* (long cucumber); a variety of cucumber with a length of 2-3 fts., used in the same way.

134. c. Cucumis sativus, *L.*, var., Jap. *Shiro-kiuri* (white cucumber); a variety of cucumber (134) with less protuberances and is of a better quality.

135. Cucumis flexuosus, *L.*, Jap. *Marudsuke, Kata-uri, Tsuke-uri*; a melon resembling Cucumis melo, *L.* (247), but green and hard. It is preserved in salt as the Japanese pickles. It is cut long in the form of a screw, dried, and preserved and is called *Kaminari-boshi*.

135. b. Citrullus edulis, *Spach.* (Cucurbita citrullus, *L.* et *Th.*), Jap. *Suikwa*, the water-melon; when the water-melons (245) in their young state have attained the size of 3 or 4 inches they are preserved in salt and eaten.

136. Cucurbita longa, Jap. *Yugao, Naga-fukube*; an annual cultivated cucurbitaceous plant with a oblong oval melon. It is 2-3 fts. in length. It is eaten either boiled or dried after being cut into pieces, being soft and sweet. Its full grown hard shells are made into vessels like gourds.

136. b. Cucurbita, Jap. *Maru-yugao, Fukube*; a variety of the former, but the melons are large and round. They are principally used to make *Kampio* by drying after cutting into long slices.

137. Cucumis, Jap. *Hime-uri, Mikan-uri*; a melon allied to Cucumis melo (247). It has the size of a swan's egg, and is eaten either raw or boiled.

137. b. Luffa petola, *Scr.*, Jap. *Hechima, Ito-uri*; the melon when green is eaten either boiled, baked, fried, or preverved in salt. (see No. 305).

137. c. Momordica charantia, *L.*, Jap. *Tsuru-reishi, Niga-uri*; this melon when green is eaten fried or roasted after having been cut into fine slices. It has a slight bitter taste. (see No. 249).

137. d. Trichosanthes japonica, *Regel*, Jap. *Kikarasuuri;* this small melon is eaten when young either boiled or preserved in salt. It has a bitter taste.

138. Solanum melongena, *L.*, Egg plant, Jap. *Nasu, Nasubi*; an annual plant of the order Solanaceae comprising many varieties. It attains a height of 2-3 fts. Its young fruits are eaten either boiled, roasted, or fried. They are also preserved in salt mixed with bran. The fruits simply preserved in salt are good to eat when boiled. It is also cut into slices and preserved in a dry state. The fruits of this plant are generally of an oval form, but some are slightly flattened at the bottom with longitudinal wrinkles near the calyx and are called *Kinchaku-Nasu* (pulse egg plant). The large kind brought from China is about 9 inches in diameter and is of a light purple colour.

138. b. Solanum melongena, *L.*, var., Long egg plent, Jap. *Naganasu;* a variety of the former with long and slender fruits. A Chinese kind grows to a length of about 2 fts. and a diameter of about 1¼ inches. It is soft and good to eat when boiled, or it can also be preserved in salt.

139. Solanum melongena, *L.*, var., Green egg plant, Jap. *Ao-nasu;* a variety of egg plaut (138) with a green rind. It is of 2 kinds, rouud and club shaped. The large round kind is inferior in taste to the long club shaped one which is called *Ao-naga-nasu* (Long green egg plant).

139. b. Solanum melongena, *L.*, var., Jap. *Gin-nasu, Tamago-nasu;* a kiud of egg plant with the fruit resembling a hen's egg. It can be eaten boiled, but is rather insipid.

139. c. Lycopersicum esculentum, *Wall.*, Jap. *Sangoji-nasu, Aka-nasu;* an annually cultivated plant of the order Solanaceae, with its fruit shaped and coloured somewhat like the persimmon. It was at first used for an ornamental purpose on account of its beautiful form, but as foreign cooking now prevails in this country it is used for culinary purposes. The fruits are eaten raw or dressed with vinegar and salt. They are also boiled, baked or reduced to paste. Its young leaves are salted and eaten

Note.—Besides those mentioned above the Ipomaea bona-box (886) and Capsicum longum (167) are to be put in this chapter.

CHAPTR VI.—EDIBLE FUNGI.

The fungi growing wild on mountains and hills or in woods are innumerable, but those used for culinary purposes are of a very limited number. The one extensively raised by artificial cultivation is *Shii-take* (Agaricus sp.). This is eaten boiled fresh, dried, or preserved in salt.

140. Agaricus sp., Jap. *Shiitake;* a fungus growing on rotten or felled woods of Quercus cuspidata, *Thunb.* (225), Q. glauca, *Thunb.*, forma sericea (565), Q. crispula, *Bl.*, (563), Q. serrata, *Thunb.*, (295), Carpinus laxiflora, *Bl.*, (567. c.) in spring, summer, and autumn. This fungus is principally propagated by artificial cultivation. The fresh ones are eaten boiled, but are generally used for commercial purposes. They are dried in two ways; one is done by exposing them in the sun, and the other by baking them.

141. Armilaria edoides, *Berk.* Jap. *Matsu-dake;* a terrestrial fungus growing under red pine trees (Pinus densiflora). A large one measures 4-5 inches in diameter, and its pileus about the same length. It is of a white colour with a strong aromatic flavour. It is wholesome and eaten either boiled or roasted when fresh. It is also preserved in cans in salt or sugar. When boiled and dried it serves for many uses.

142. Agaricus sp., Jap. *Shimeji;* a fungus growing on the ground in woods in autumn. It consists of different varieties, but the ordinary kind is of a white or grayish colour. The diameter of its pileus is about 3-4 inches, and its stipe is about 4 inches. It is eaten by boiling; it is also dried or preserved in salt.

143. Agaricus sp., Jap. *Hatsu-dake;* a fungus growing on the ground in woods in summer and autumn. It is concaved on the upper side of the pileus and resembles a small Japanese wine cup. There are two kinds, red and green; the former is called *Aka-hatsutake* and the latter *Ao-hatsutake*. Both are eaten roasted or boiled when fresh.

144. Agaricus sp., Jap. *Samatsu-dake;* a fungus growing in pine woods in summer. The form and size resemble those of *Matsu-take* (141), but inferior in taste. It is esteemed for its early production.

145. Agaricus sp., Jap. *Sembon-shimeji;* a fungus growing in tufts on the ground. It has a delicious taste. Besides this there is a kind resembling *Shimeji* (142) with slender stipes growing also in tufts, but is different, though it has the same name.

146. Agaricus sp., Jap. *Kuri-take;* a fungus growing on decayed chestnut trunks. Late in autumn it is taken and eaten by boiling.

146. b. Agaricus sp., Jap. *Mai-take;* a fungus growing in tufts on the bark of rotten trees. It is eaten fresh by boiling and preserved after drying.

147. Exidia auricula, *Juda.*, Jap. *Kiku-rage;* a fungus growing on the bark of decayed trees. It is like a man's ear with a diameter of 3-4 inches and of a brown colour. On mountains it is often seen growing on the rotten part of several kinds of trees, but that growing on Lambucus racemosa, *L.*, var. sieboldiana, *Miq.* is esteemed as the best. It is dried, preserved, and used as a vegetable. When eaten it makes a noise as *Kurage* (a kind of medusa).

148. Lichenes digitatus, *Ach.*, Jap. *Iwa-take, Iwa-goke;* this is not a fungus, but is put in this section on account of the similarity of its use with that of the fungus. It grows on rocks among mountains. Its upper surface is flat, smooth, and of a grey colour, but the under-part is black, rough, and provided with short stipes. It is dried, preserved, and used as food.

149. Tuber spadiceum, Jap. *Shōro ;* a terrestrial fungus growing in sandy soil near the sea-shore and also among pine trees in spring and summer. Its shape is a small round ball about ½ or 1 inch in diameter with somewhat of a pine resinous flavour. It is divided, into 3 kinds according to its colour, namely *Kome-shoro, Awa-shoro,* and *Hiye-shoro.* It is used in cooking when fresh and also preserved in salt or sugar.

150. Hydnum wrightii, Jap. *Kawa-take, Shishi-take, Kō-take ;* a terrestrial fungus growing in mountainous regions on heaps of fallen leaves under trees. It is like a small shallow cup about 5 or 6 inches in height and covered with scaly hair. It is used in cooking when dried. It is of a dark colour with a nice flavour.

150. b. Hydnum wrightii, Jap. *Rōji, Rōjin ;* a fungus resembling very much the former. It is dark on the outside and white inside. It has a slight bitter taste and is eaten when roasted. It is like a Japanese umbrella in form.

CHAPTER VII.—EDIBLE ALGAE.

This Chapter includes the aquatic plants of the order Algae. They are very numerous, but those for economic purposes are few in number. Those described in this section are dried and used for food. Some others used as starch are described in the chapter of " Different uses." Some of the algae are raised by artificial cultivation.

151. Porphyra vulgaris, *Sur.,* Jap. *Asakusa-nori, Ama-nori ;* this is an algae growing on rocks where the sea is shallow, but it is also cultivated artificially by placing branches of trees in the mud of the sea which enables it to grow plentifully. In winter and spring it is taken, dried, and used for food called *Asakusa-nori.* The eastern provinces are noted for this production. The products in different places are almost the same in all respects, but *Kuro-nori* (black algae) and *Upurui-nori* are somewhat different in shape and colour.

152. Alaria (Ulopteryx) pinnatifida, Jap. *Wakame;* an algae growing in the sea. Its stem is the length of 3-4 fts. expanding into a leaf parted into many divisions at the top. Late in spring its young leaves are taken, dried, and preserved. They are eaten by soaking in vinegar, roasting, or boiling.

The *Ito-wakame* of *Ise* province, *Naruto-wakame* of *Awa* province, etc. are noted products of different places. The *Nanbu-wakame* has long segments.

From the ear-like folds attached to both sides of the stem an elastic glue called *Wakame-tororo* is taken and eaten.

There are also *Ao-wakame* (green *wakame*) and *Hira-wakame* (flat *wakame*) with entire leaves.

153. Laminaria japonica, *Aresch.*, Jap. *Konbu, Kobu, Hirome;* a large long algae growing in the cold seas of *Hokkaido* and the nourthern provinces. This is taken in summer and is preserved by drying. It comprises many subspecies different in form, taste, and colour.

A kind called *Atsu-konbu* (thick Laminaria) has a considerable breadth and is used to make *Hana-ori-konbu* and *Moto-soroye-konbu*. Its length is about 6 or 7 fts. and is of a good thickness. It is delicious and is used for cooking.

Another kind called *Mitsu-ishi* has a length of 3-4 fts. and a breadth of 3-4 inches, and has a good taste.

Naga-konlu (long Laminaria) has an extensive length of 60-70 fts. and a breadth of 5-6 inches, and is used to make what is called *Naga-kiri-konbu* (long cut *Konbu*) which is much exported to China.

Kuro-konbu (black Laminaria) is small in size, of a dark colour, and inferior in taste to the preceding. *Konbu* tinted with verdigris is called *Ao-ita-konbu;* when cut into fine pieces it is called *Kizami-konbu*.

Konbu (Laminaria) is eaten boiled, roasted, fried, or preserved in salt or sugar.

It is used to put in boiled rice after having been cut into small pieces. It is also used to give a flavour to soup, or as an infusion like tea.

The Chinese call the Laminaria *Kaitai*, and the cut one, *Taishi;* both are esteemed by them as a delicious food.

153. b. **Laminaria** sp., Jap. *Hosome, Bonme;* a small kind of Laminaria inferior in taste is produced in the seas of the northern provinces. It is used at the feast of lanterns called *Bon* in Japan, whence derived the name *Bonme*.

153. c. **Laminaria** sp., Jap. *Hokka-kombu;* a species of Laminaria (153) growing in the sea of *Rikuzen* or *Rikuchiu*. It is thin and inferior in taste.

153. d. **Laminaria** sp., **Arthrothamnus bifidus,** *Jag.*, Jap. *Neko-ashi-konbu, Mimi-konbu;* a species of Laminaria (153) growing in the cold seas of *Nemuro* and *Kushiro* and their neighbouring provinces. It is about 4 fts. in length and 2-2½ inches in breadth. As it has ear-lobe-like protuberances at both sides of the base of the frond, it is called ear-like or cat's foot Laminaria and has a good taste.

153. e. **Laminaria** sp., Jap. *Tororo-konbu, Chizimi-konbu;* a species of Laminaria (153) growing in the seas of *Nemuro* and *Kushiro* in *Hokkaido*. It has a length of 3-4 fts. with a breadth of about 2 inches, covered with wrinkles on the whole surface. It is very rich in a gluey fluid and is eaten like the gruel of the dioscorea tuber.

154. **Capea elongata,** *Ag.*, Jap. *Arame, Kurome;* an algae much produced in the seas of different provinces. It is divided into parts containing several leaves at the top of a long stem; each leaf has a length of 1-2 fts. with a breadth of 1½-2½ inches and is flat in form and of a dark grey colour covered with wrinkles, but when dried it becomes quite black. They are gathered late in spring and preserved by drying and used as food.

154. b. **Capea richardiana,** Jap. *Kajime, Sagarame;* an algae resembling very much Capea elongata, *Ag.* (154) in form,

but narrower. It has rough longitudinal wavy wrinkles. The taste is almost the same, but inferior to the preceding. Its stem is strong with a length of about 2 fts. and used as walking sticks and handles of umbrellas, and for other similar purposes.

155. **Chondria**, Jap. *Hijiki;* an algae growing on rocks in shallow seas. It is 3–4 inches in length having leaves and branches. When fresh it is dark green, but becomes black when dried. It is gathered in spring, dried, and eaten after boiling in water. The product in the province of *Ise* is noted as the best kind.

156. **Chondria**, Jap. *Naga-hijiki, Chōsen-hijiki, Michi-hijiki;* a species of the *Hijiki* (155) with a length of about 1 ft. It is of the same quality as the former.

157. **Enteromorpha compressa**, *Grev.*, Jap. *Ao-nori;* a fine algae growing on rocks and woods in water near the mouths of rivers. It has fine fibres. In winter and spring they are gathered when they have grown to the length of 3–4 inches. They are preserved by drying and eaten by baking, being esteemed for their flavour.

158. **Phicoseris smaragdina**, *Kg.*, Jap. *Aosa, Tisa-nori;* a broad flat green algae growing on stones or woods in shallow sea water. From winter to spring it is taken, dried, and preserved. It resembles the preceding in taste, but inferior. One kind called *Bekko-aosa* is very pretty on account of its lustrous green colour.

158. b. **Phicoseris australis**, *Kg.*, Jap. *Kawa-nori;* an algae growing on stones in streams among valleys. Its form and colour resembles those of Phicoseris smaragdina, *Kg.* (158). In spring months it is collected, dried, preserved, and eaten when roasted. It is superior in flavour to the latter (158).

Those produced in a river *Daiya-gawa* in *Nikko* and a river *Shiba-kawa* at the foot of *Fuji* mountain are esteemed as the best quality.

158. c. **Phylloderma sacrum**, Jap. *Suijenji-nori;* an

algae growing in the stream of the valley of the temple *Suijenji* near *Kumamoto* in the province *Higo*. It is soft and of a dark green colour and of different sizes. It is preserved by drying in the form of paper, and eaten boiled or dipped in vinegar. The prepared algae called *Jusentai* of the province *Higo* is of the same quality.

159. **Gelidium corneum**, *Lamour.*, Jap. *Tokorotengusa*, *Tengusa ;* a finely branched algae growing on stones in sea-water. It is about 5-6 inches in length and dark purple in its colour when fresh, but turns yellow when bleached and dried. It is made into a jelly by boiling. This jelly when dried and congealed is called *Kanten* (gelatine vegetale in French or Japanese isinglass in English). It is also made to *Kanten-gami* (gelatine vegetale in the form of paper), *Mishima-nori*, etc.

159. b. **Campylaephora hypnaeoides**, *J. Ag.*, Jap. *Yego*, *Ukeudo*, *Magari ;* a parasitic algae growing on other sea-weeds. It is very fine and divided into many branches provided with hooks at their ends which readily entangle thermselves with other objects. It is dark purple when fresh, but turns white when bleached. It is eaten by reducing to a gelatinous substance by boiling, or used to mix with the Japanese isinglass or gelatin of Japan. The whole plant tinted red is called *Shōjō-nori* and is used as an ornament.

159. c. Jap. *Igisu ;* an algae growing on stones in the sea with numerous fine branches. It is dark purple when fresh, but turns white when bleached. It is eaten in a gelatinous state.

160. **Halochloa macrantha**, *Kg.*, Jap. *Hondawara*, *Kawaramo ;* an algae growing on rocks in the sea. It grows to a length of 2-3 fts. with alternate leaves which are provided with numerous small air-cells.

It is eaten when young, and also used as an ornament for new year's day.

160. b. **Mesogloia decipiens**, *Sur.*, Jap. *Mo.tsuku ;* an algae growing on Halochloa macrantha, *Kg*. It is of a smooth

and soft nature having numerous fine branches and is eaten preserved in salt.

161. **Hallymenia dentata**, *S. Z.*, **Gleopeltis rigens,** *Grev.*, Jap. *Tosaka-nori;* a thick, broad, and pink algae with dentate edges resembling a cock's comb attaining a length of several inches. There is a kind of a thinner and softer nature having many parts. They are preserved by drying and are sometimes eaten boiled or in a state of jelly.

161. b. **Gracilaria conferioides**, *Grev.*, **Gigartina tenelle,** *Harvey*, Jap. *Ogo, Ogo-nori;* a long fine algae divided into many branches, growing on stones or shells in muddy sea-water. It attains a length of about 2 fts. In the eastern provinces they are used to ornament the table by placing them beside the dishes. They are boiled in lime water to make hard and stiff. They are also used to make glue when dried.

Note.—Besides those mentioned above there are many other algae which are eaten, but they are beyond description in this limited volume; for example *Miru* (Codium), *Shiramo* (Sphaerococcus japonicus, *Sur.*), *Umi-sōmen* (Nemalion vermiculare), *Tsurumo, Matsumo, Kyono-himo* or *Kawagishi, Kome-nori*, etc. are used for this purpose. Also *Tsuno-mata* (Gymnogongrus japonicus, *Sur.*) and *Funori* (Gigartina intricata, *Sur.*) described in the division of plants of "Different uses" and some of the fresh water algae are eaten in the same way.

CHAPTER VIII.—CONDIMENTS AND SPICES.

This Chapter includes the plants which have an aromatic flavour and pungent taste increasing appetite. Some of their leaves are used as culinary vegetables, and some of the seeds are used as medicine.

162. **Eutrema wasabi**, *Maxim*, Jap. *Wasabi;* a perennial herb of the order Cruciferae growing wild in valleys, but often cultivated near streams and rivers. The roots are used as a stimulant, and the leaves and stems as a vegetable.

163. **Raphanus sativus**, *L.*, var., Jap. *Nedsumi-daikon*, *Karami-daikon;* a biennial cruciferous vegetable which is the celebrated product of the village *Uyeno* at the foot of *Ibuki-mountain* in the province *Omi*. It is short and thick at the end in the form of a club. As it is provided with rat's tail like taper roots, it is called *Ibuki-daikon* or *Rat-daikon*. It is very acrid in taste and used as a condiment, but it is also good to be eaten boiled.

Besides this, *Sangatsu-daikon* (March-raddish), *Natsu-daikon* (Summer raddish), etc. are used for condiments.

163. b. **Sinapis cernua**, *Thunb.*, Jap. *Karashi;* the seeds of Sinapis cernua, *Thunb.*, are grind into powder and used as a condiment or preserved in salt.

164. **Citrus aurantium**, *L.*, Jap. *Yudsu*, *Mochi-yudsu;* an evergreen cultivated tree of the order Aurantiaceæ. It attains a height of 10-15 fts. It thrives in cold regions as well as in warm countries. Early in summer it produces flowers and gives fruits in winter. It is the size of a wrist and of a pure yellow colour when fully ripe. Its rind is very fragrant and is eaten fresh, boiled, or preserved in sugar. Its flower-buds and young fruit-rinds are used in cooking to give the food an aromatic flavour.

164. b. **Citrus aurantium**, *L.*, var., Jap. *Toko-yudsu*, *Hana-yudsu;* a variety of the preceding (164); the smaller fruits remain on the branches for a long time. It is inferior in quality, but of the same use. It is usually used when young.

165. **Zanthoxylum piperitum**, *D. C.*, Jap. *Sansho;* a wild mountain deciduous shrub of the order Zanthoxylaceæ. It is also cultivated in gardens. Its ripe fruits, young flower-buds, and leaves, as well as the inner bark of the stem which is called *Kara-kawa* are eaten when boiled. A kind called *Asakura-sansho* has shorter thorns, but the fragrance of its leaves and fruits is stronger.

165. b. **Prunus Grayana**, *Max.*, Jap. *Uwamidsu-sakura;* the fruits of this tree are called *An-nin* in *Yechigo*. The flower-

buds and young fruits are eaten when preserved in salt and have a pungent delicious taste.

166. Phellopterus littoralis, *Fr.*, Jap. *Yaoya-bōfū, Hama-bōfū;* a triennial umbelliferous herb growing in sandy ground near sea shores. Its young soft leaves are eaten raw. In *Tokio* they are cultivated in farmyards; they grow throughout the year, and their young leaves are used to decorate dishes.

167. Capsicum longum, *L.*, Jap. *Tōgarashi, Nanban;* an annual cultivated plant of the order Solanaceae comprising many varieties. The fruits of the ordinary kind are red in colour, but some are yellow and others dark purple. In form some are long and thin, and others round or ovate. The plant with long fruits is called *Nikko-tōgarashi*, and the short one, *Taka-no-tsume*. Both are very acrid and hot, but there is a kind called *Ama-tōgarashi* which owing to the mildness of its acridity is eaten as a vegetable. The variety here mentioned is called *Tenjiku-mori* or *Yatsu-busa*, and is much cultivated in the vicinity of *Tokio*.

168. Capsicum longum, *L.*, var., Jap. *Shishi-tōgarashi;* a variety of Capsicum longum. *L.*, with wrinkles on the skin. It has the same use as the preceding.

169. Capsicum cerasiforme, Jap. *Yenomi-tōgarashi;* a kind of pepper with its fruits resembling those of Celtis sinensis. They have the same quality and use as the pepper.

170. Perilla arguta, *Benth.*, Jap. *Shisō;* an annual cultivated plant of the order Labiatae growing to a height of about 2 fts. Its young seeds are eaten raw or boiled. Its leaves and flower-racemes are used as condiments or preserved in salt; especially the leaves are used to give a purplish red tint to the salt-preserved Prunus mume. A variety with wrinkled leaves has a deep purple colour. In early spring the seeds are sown under glass and their cotyledons are used as a condiment.

171. Perilla arguta, *Benth.*, var., Jap. *Aoso, Shiroso;* a variety of Perilla arguta, *Benth.*, with green leaves and stems and white flowers. As it has a strong flavour it is used as a spice or preserved by drying or in salt.

172. P olygonum maximowiczii, *Regel.*, Jap. *Yanagi-tade;* an annual cultivated plant of the order Polygonaceae. There are several varieties, some with narrow, and others with broad leaves, which are purple or green. The kind here mentioned is the green narrow leaved variety, and as it has a sharp acrid property its young leaves are used in cooking. A kind called *Kawa-tade* thrives well even in winter.

173. Polygonum maximowiczii, *Regel.*, var., Jap. *Kinshi-tade;* a variety of the former with fine narrow leaves. There is a kind with purple leaves. They are also the same in quality and use.

173. b. Actinidia polygama, *Planch*, Jap. *Matatabi;* a deciduous climbing plant of the order Dilleniaceae growing wild on mountains. In summer it bears white flowers resembling those of Prunus mume. Its leaves are eaten boiled, and also its young fruits are eaten after being salted. Both have an acrid taste. Cats are very fond of this plant.

174. Zingiber officinale, *Roscoe*, Jap. *Shōga, Haji-kami;* a perennial cultivated plant of the order Zingiberaceae attaining a height of about 2 fts. The young shoots come forth from the new roots produced from the old stocks. They are very ornamental. Their red stems have an agreable aromatic flavour and a slight acrid and pungent taste. They are used for various purposes in cookery. Its roots have a strong acid taste. They are used as a condiment, and are also preserved in salt, sugar, or syrup. They thrive well in warm regions where the roots are sound and large with a good aromatic taste, while those cultivated in cold regions are small and hard with numerous fibres.

Note.—In the division of Leaf-vegetables Oenanthe stolonifera, *D.C.* (58), Aralia cordata, *Th.* (61), and the flowers of Petasites japonicus, *Th.* (67) and Allium fistulosum, *L.* (80); in the division of Flower-vegetables the flower of Zingiber Mioga, *Roscoe*, and its young shoots, young plants of Mentha arvensis, *L.*

var. vulgaris, *Benth.* (446), rind of citrus nobilis, *Lour.* (230), and seeds of Sesamum indicum, *L.* (309), Cannabis sativa, *L.* (323), etc. are used as condiments and spices.

CHAPTER IX.—FRUITS.

This Chapter includes the fruits produced from trees and herbs.

There are several kinds. Most of them are eaten raw, but many are preserved being dried or kept in salt or sugar, and some are used for fermenting wine. Many plants of this division furnish timbers, but they are not mentioned here.

175. Prunus mume, Sieb. et Zucc., Jap. *Mume, Ume;* a deciduous tree of the order Rosaceae attaining a height of about 10 fts. It bears flowers early in spring before the leaves appear. It comprises many different varieties, being single or double petaled, and pink or white coloured. The fruits also differ in size. These various kinds are cultivated more for the sake of their flowers than for their fruits. Its fruits are gathered before being fully ripe and preserved in salt. They give a red tint when mixed with the leaves of Perilla arguta, and are eaten as a relish. It is also used for various preserves, such as *Mume-bishio* (jelly), dried plum, etc.

176. Prunus mume, *S.* et *Z.*, var., Jap. *Yatsubusa-no-mume;* a variety of Prunus mume, *S.* et *Z.* It has about 8 fruits on one calyx, but as some fall off before they ripe only two or three come to maturity. They are not very good to eat, but prised rather for curiosity.

177. Prunus mume, *S.* et *Z.*, var., Jap. *Bungo-mume;* a variety of Prunus mume, *S.* et *Z.* (175) with a larger fruit about 2½ inches in diameter, but not so prolific as the common mume. They are eaten raw, boiled, or preserved in salt or sugar.

178. Prunus mume, *S.* et *Z.*, var., Jap. *Komume, Shina-no-mume;* a variety of Prunus mume, *S.* et *Z.* (175) with very small furits which ripen early in June. They are noted for their small size and are preserved in salt or sugar.

179. Prunus mume, *S.* et *Z.*, var., Jap. *Tokomume, Aomume, Fudan-mume;* a variety of Prunus mume, *S.* et *Z.* Its fruits remain on the tree long after maturity, whence they are called *Toko-mume* (everlasting plum) and are eaten fresh. They may be kept a long time without decay.

180. Prunus armeniaca, *L.*, Jap. *Andsu, Karamomo;* a deciduous tree of the order Rosaceae resembling Prunus mume, *S.* et *Z.*, in form. It attains a height of about 10 fts. In spring it blooms next to P. mume, *S.* et *Z.*, with single pinkish white flowers. It ripens early in summer. Its fresh is easily separated from the seeds. They are yellow when ripe and of a delicious sweet flavour. They are eaten either raw or dried.

181. a. Prunus triflora, *Roxb.*, Jap. *Sumomo, Su-ume;* a deciduous fruit tree of the order Rosaceae attaining a height of about 10 fts. The white single flowers appear in spring after Prunus mume, *S.* et *Z.* The round and lustrous red fruits ripen in summer and are delicious to eat. They are also picked before maturity and preserved in salt. There are several varieties.

181. b. Prunus triflora, *Roxb.*, var., Jap. *Urabeni-sumomo;* a variety of the preceding with a deep red pulp. In the provinces of *Kiushiu* it is named *Ikuri*. There is one variety with good round fruits in the province of *Satsuma*, being called *Yonemomo;* it has also a red pulp.

182. Prunus triflora, *Roxb.*, var., Jap. *Shiro-sumomo;* a variety of P. triflora, *Roxb.* (181), differing from it only by its yellowish white colour. Another variety of a yellow colour is also called *Shiro-sumomo* and is superior in taste.

183. Prunus triflora, *Roxb.*, var, Jap. *Togari-sumomo, Hadankio;* a variety of Prunus triflora, *Roxb.*, with large and pointed fruits. There are two kinds, one red and the other white. Another variety with round fruits is calld *Botankyō*.

184. Amygdalus persica, *Benth.* et *Hook*, Jap. *Momo;*

a deciduous tree of the order Rosaceae attaining a height of about 10 fts. In spring it blooms pretty flowers of various kinds, single or double, red or white, etc. Delicious fruits are produced from those of the single pink flowers. The fruits ripen either in summer or in autumn. The size is about 1-2 inches. It is good to eat fresh, or preserved in sugar or salt.

184. b. Amygdalus persica, *Benth.* et *Hook.*, var., Jap. *Kam-momo, Fuyu-momo;* a late ripening variety of peach (184). The fruits are preserved till winter and even to spring.

185. Amygdalus, Jap. *Amendo;* a dwarf variety of peach (184) attaining a height of 60-70 fts. When it is about a foot high it forms a pretty looking dwarf plant with many branches covered with long narrow leaves and flowers in clusters. There are several kinds of flowers, single or double, pink, white, or variegated, etc. Its fruits are ripe in autumn.

186. Amygdalus, Jap. *Dsubai-momo, Tsubaki-momo;* a variety of peach (184) with red smooth delicious round fruits. There is one which does not ripen to red; it is called *Aodsubai* (green variety).

186. b. Myrica rubra, *Sieb.* et *Zucc.*, Jap. *Yama-momo;* this tree yields great quantities of fruits in warm regions. They are round about the size of a thumb. When ripe they are of a dark red colour and rich in a sweet juice. There are the varieties with white and yellow fruits.

187. Prunus tomentosa, *Thunb.*, Jap. *Yusura-mume;* a deciduous shrub of the order Rosaceae attaining sometimes a height of 7—8 fts, but generally smaller and slender than the preceding. In spring it produces five petaled white flowers before the leaves. In summer it bears round dark red lustrous fruits which resemble cherries. They are sweet, juicy, and delicious.

188. Zizyphus vulgaris, *Lam.*, Jap. *Natsume;* a deciduous tree of the order Rhamnaceae attaining a height of about 20fts. In the beginning of summer it blooms small yellowish green flowers on the branches. In autumn its oval or oblong fruits are yellow when ripe, but gradually turn to a reddish brown colour afterwards. It is eaten fresh or when dried and preserved.

188. b. Zizyphus vulgaris, *Lam.*, var., Jap. *Naga-natsume*, *Tokuri-natsume;* its fruits are long and pointed when ripe.

189. Pyrus communis, *L.*, Pear, Jap. *Nashi*, *Ari-nomi;* a deciduous tree of the order Rosaceae attaining a height of about 30 fts., but in cultivating it is generally trained down over trellis. In spring, it blooms single white flowers, before it sprouts. The fruits ripen in summer; they are of different forms, sizes, and tastes. They are eaten fresh, and preserved by drying or made into jam. The variety drawn here is common called *Taihei* in Tokio. The fruits, which ripen late, are preserved till the next summer. Many varieties are cultivated in different places. We will mention here few varieties in the following lines.

189. b. Pyrus communis, *Lam.*, var., Jap. *Ao-nashi;* its green fruits ripen early and are juicy.

189. c. Pyrus communis. *Lam.*, var., Jap. *Inu-nashi*, *Yama-nashi;* an original species of pear attaining a considerable height, with small hard fruits and thorns on the branches. The seeds are sown and the young plants are used for grafting stocks.

189. d. Pyrus communis, *Lam.*, var., Jap. *Tane-nashi-arinomi*, *Tane-nashi-innashi;* the fruits are very small, but have no seed.

190. Pyrus ringo, Jap. *Ringo;* a deciduous tree of the order Rosaceae, attaining a height of about 10fts., with slender extended branches. In spring it blooms after producing the leaves. The buds are pink, but when open, they are almost white. The fruits are round about an inch in diameter, and their parts facing to the sun are pink. They are eaten fresh, and may be preserved by drying after cutting into slices. A variety with very small aciduous fruits is called *Ko-ringo*.

190. b. Pyrus ringo, var., Jap. *Beni-ringo, Rinkin;* the trees (635) produce plenty of fruits in cold regions. The fruits are round or oval, with a diameter of about an inch. In autumn they ripen and are scarlet. Their taste is better than that of the former. They are eaten fresh and also preserved by drying.

191. Pyrus chinensis, *Pair*, Jap. *Kwarin, Karanashi;* a deciduous tree of the order Rosaceae, attaining a height of 20—30fts. Its bark peels down itself every year, and cloud-like variegated scars are left behind. Late in spring it produces single pink flowers with the leaves. Its fruits ripen late in autumn. They are yellow and oval with rough surfaces. They are too sour to be eaten fresh; so they are baked or steamed.

191. b. Pyrus cydonia, *L.*, Jap. *Marumero;* a species closely allied to the preceding, attaining a height of 70—80fts., with many branches growing in clusters. The flowers are pink and about an inch in diameter. The fruits are covered with fine hair, and their surfaces are very uneven. They are about $2\frac{1}{2}$ inches in length, and are eaten fresh. This species thrives better in cold climates.

191. c. Pyrus japonica, *Th.*, var. genuina, Jap. *Boke, Karaboke;* it produces many fruits, which resemble those of Pyrus chinensis, Pair, but smaller in shape and inferior in taste.

191. d. Pyrus japonica, *Th.*, var. pygmaea, Max., Jap. *Noboke, Kusaboke, Shidomi;* this tree (633) produces many fruits which resemble those of Pyrus japonica, *Th.* (632). The fruits are round with uneven surfaces, and are very sour in taste.

192. Photinia japonica, *Th.*, Jap. *Biwa;* an evergreen tree of the order Rosaceae, attaining a height of about 20fts. In early winter, it blooms single, white, and fragrant flowers disposed in panicles, and produces fruits in the next summer. The fruits are yellow and round, covered with fine hair. They are as large as finger-heads, and are very delicious and aciduous; so they are highly prized. They contain 2—3 large seeds. A variety, with the fruits which skin has white powder, is called *Shiro-ko-biwa*.

192. b. Photinia japonica, *Th.*, var., Jap. *Tōbiwa;* its fruits are large, with a very good taste, and the leaves are also large. When 8—9 years are passed after the seeds were sown, the young trees grow 6—7 fts. high, and produce very good fruits as their mother trees.

192. c. Photinia japonica, *Th.*, var., Jap. *Nagatōbiwa;* its fruits are oval and sometimes obovate.

193. Crataegus cuneata, *S.* et *Z.*, Jap. *Sanzashi;* a deciduous shrub of the order Rosaceae, attaining a height of 5—6 fts., with many thorny branches. In spring it produces single white flowers in clusters, being followed with round red or yellow skinned fruits about ⅜ inch in diameter. The fruits are slightly sweet and acidouous.

193. b. Crataegus sanguinea, *Pall.*, Jap. *Ōsanzashi;* a species closely allied to the preceding, with larger leaves and fruits, attaining a height of about 10 fts.

194. Diospyros kaki, *L.*, Persimmon, Jap. *Kaki;* a deciduous tree of the order Ebenaceae, attaining a height of 20—30 fts. In late spring, it shoots forth new branches and leaves; in early summer, it opens male and female flowers separately; and in autumn, its fruits ripen and are yellowish red or crimson. The forms of the fruits are various, and their taste is sweet or astringent. *Kizawashi* (sweet *kaki*) is eaten fresh, and *Shibu-kaki* (astrigent *kaki*) is made into *Umi-kaki*, *Sawashi-kaki*, *Shibunuki-kaki*, *Amaboshi-kaki*, *Koro-kaki*, *Kaki-tsuki* etc., and then edible. The variety drawn here is a *Kizawashi* called *Zenjimaru* being produced abundantly in eastern provinces. The varieties of this plant are very numerous; so only a few among them are described in the following articles.

195. Diospyros kaki, *L.*, var., Jap. *Hachiya-gaki; Mino-gaki;* a variety with large oblong fruits, being about 3½ inches in height and 8—9 inches in circumference. The seeds are long and narrow, and few in number. The fruits are very good to eat when made into *Amaboshi-kaki* or *Koro-kaki*. They are chiefly produced in *Hachiya*-village of *Mino*-province, whence the name is derived. They are cultivated everywhere, but are different more or less from each other. The one called *Fujiyama* belongs also to this variety.

195. b. Diospyros kaki, *L.*, var., Jap. *Saijo-gaki;* a variety with oval and middle sized fruits, much cultivated in the central provinces. They are made into *Shibu-nuki-kaki*.

196. Diospyros kaki, *L.*, var., Jap. *Yemon-gaki;* a variety with flat and somewhat square fruits. Some largest fruits are about 2½ inches high and 9—10 inches in circumference. They lose their astringent taste when put in an empty *sake*-tub. They are juicy and are considered as the best *Shibunuki-kaki*. The fruits of a variety called *Yama-yemon* are flatter and depressed at the heads.

197. Diospyros kaki, *L.*, var., Jap. *Hyakume-gaki;* when its round large fruits fully ripen, they weigh a hundred *momme* ($\frac{3}{6}$ lb.) and are red, with black cloud-like figures at the heads. They are few in seeds, and are edible fresh.

197. b. Diospyros kaki, *L.*, var., Jap. *Gosho-gaki;* a variety of *Kizawashi* with flat and somewhat quadrangular prismatic fruits which turn red when ripe. It is the best species of persimmons, being very pretty in appearance, delicious in taste, small and few in seeds, and preservative for a long time. It was a famous product of *Gosho*-village of *Yamato*-province, whence the name derived, but now it is planted everywhere.

197. c. Diospyros kaki, *L.*, var., Jap. *Hi-gaki* (Drying persimmon); its fruits remain withered on the branches, even after ripen.

197. d. Diospyros lotus, *L.*, Jap. *Shinanogaki, Mamegaki;* a species of persimmon with smallest fruits, being as large as fingerheads. They are round or oblong in form. They are gathered in winter and dried to eat. As the preceding, they also dry up on the branches. From the young fruits *Kaki-shibu* (an astringent juice) is obtained.

198. Punica granatum, *L.*, Jap. *Zakuro, Jakuro;* a deciduous shrub of the order Myrtaceae, attaining a height of 8-9 fts., growing in group, but sometimes being tall as a large tree. The flowers bloom during the rainy season of summer. There are many varieties with single or double, and light red, dark red or white variegated flowers, but only the variety with single and dark red flowers produce fruits in late autumn. The fruits are light red and round, and have thick skin with sepals at the top. When ripe, the skin bursts and exposes red seeds with a pellucid pulpy coating. There are two sorts, one with aciduous and the other with sweet pulp.

198. b. Punica granatum, *L.*, var., Jap. *Shiro-zakuro;* its pulp is almost white, slightly shaded with pink, and the taste is very delicious.

199. Vitis vinifera, *L.*, Grape, Jap. *Budō;* a deciduous climber of the order Vitaceae. The stems of some large vines are several inches round. It is cultivated in gardens and extended over trellis. In early summer it produces small yellowish green flowers disposed in panicles from the axils of the leaves, being succeeded with the bunches of grapes which ripen in autumn. The grapes are $\frac{1}{2}-\frac{4}{5}$ inch in size. There are at least 60-70 grapes in each bunch. The fruits are generally ovate, but some are round. The fruit-skin is green shaded with purple. The variety produced in province *Kai* is beautifully shaded with purple.

There are several other varieties with purple or white skins. They are eaten fresh, and have a sweet and refleshing taste. They are dried and preserved in sugar. Wine is made from them.

199. b. Vitis labrusca, *L.*, Jap. *Yama-budō;* a deciduous climber of the order Vitaceae, growing wild in mountains. The leaves are broad, and their under surface is covered with brown hairs. The fruits are purplish black and inferior in quality being too aciduous. There are several other wild varieties. The one called *Yebidsuru* is closely allied to this, but smaller.

200. Actinidia arguta, *Planch.*, Jap. *Sarunashi, Shirokuchi, Kokuwa;* a deciduous climber of the order Dilleniaceae growing abundantly in mountains. The largest stems of this plant are about 1½ fts. round In summer it blooms single white flowers, about ⅜ inch in diameter, being succeeded with round berries, which are eaten fresh or dried.

201. Broussonetia papyrifera, *Vent.*, var., Jap. *Himekōzo;* a deciduous tree of the order Urticaceae, attaining a height of about 10 fts. It has distinct male and female flowers upon separate trees. The fruit forms a round ball congregated of small sweet red pulpy seeds. This is closely allied to Broussonetia papyrifera, *Vent.* and *B. kajinoki, Sieb.*, occuring wild on mountains and in fields. The fruits have almost the same form.

202. Rubus parvifolius, *L.*, Jap. *Nawashiro-ichigo;* a deciduous trailing plant of the order Rosaceae growing wild in plains and bushes, attaining a length of 4-5 fts. The leaves are ternate and their under surface is white. It blooms five-petaled purplish small flowers on numerous small branches at the end of the stem. They are succeeded by red juicy sweet berries which ripen in summer. All these berries are formed by the collection of small pulpy seeds, having the same taste and different colours. These grow wild abundantly, but are also cultivated in gardens. The fruits are eaten fresh or made into jam or wine.

203. Rubus phœnicolasius, *Max.*, Jap. *Yebikara-ichigo*, *Saru-ichigo;* a species closely allied to the preceding, but of a larger form, growing wild among mountains, attaining u height of 6-7 fts. The leaves and stems are covered with red hair. The fruits are yellowish red in colour and are ripe in autumn.

204. Rubus buergeri, *Miq.*, Jap. *Fuyu-ichigo;* a deciduous small shrub of the order Rosaceae, attaining a height of 3-4 fts. and growing wild everywhere. The leaves and stems are furnished with hooked spines. The former have 5 or 7 lobes somewhat like a maple-leaf. In the beginning of summer it blooms five petaled white flowers about an inch in diameter from the axils of the leaves and yields yellowish berries.

205. Rubus palmatus, *Th.*, Jap. *Awa-ichigo;* a deciduous small shrub of the order Rosaceae. Its stem is 3-4 fts. high, growing wild everywhere. Both the stem and leaf have thorns. The leaf is like that of maple, and in early summer it opens white flowers which have five petals, being about an inch in length. Its fruits are yellow when ripe.

206. Rubus incisus, *Th.*, Jap. *Ki-ichigo*, *Niga-ichigo;* a species closely allied to the preceding, growing wild on mountains and in bushes, with smaller flowers and red berries which are of an inferior taste.

207. Rubus trifidus, *Th.*, Jap. *Kaji-ichigo*, *Chosen-ichigo;* a deciduous small shrub of the order Rosaceae, with a straight stem of a height of 5-6 fts. The leaves are broad and five lobed with the diameter of 6-7 inches. In summer it blooms five-petaled white flowers, and the sweet yellow berries are ripe in autumn.

208. Rubus morifolius, *Sieb.*, Jap. *Kuma-ichigo;* a deciduous shrub of the order Rosaceae, attaining a height of 5-6 fts. Its leaves and stems are furnished with sharp thorns. The leaves are broad and 4-5 inches in diameter. The fruits are large and red.

208. b. Debugeasia edulis, *Wedd.*, Jap. *Yanagi-ichigo*, *To-ichigo*, *Karasu-yamamomo;* a deciduous shrub of the order Urticaceae, growing abundantly in warm districts. The fruits are yellow and resemble those of the straw-berry.

209. Vaccinium vitis-idaea, *L.*, Jap. *Oyama-ringo*, *Hama-nashi;* a tiny evergreen shrub growing in alpine regions, attaining a height of 6-7 inches. In the beginning of summer it blooms small white and pink shaded flowers. The berries are round and of a red colour with an aciduous taste. They are $\frac{1}{4}-\frac{1}{3}$ inch in size, and are eaten fresh or preserved in salt or sugar.

210. Vaccinium oxycoccos, *L.*, Jap. *Tsuru-kokemono;* a procumbent tiny evergreen plant of the order Ericaceæ, growing in moist places of mountains. It has many stems which produce only one flower at each end, being succeeded with small berries which droop. Their shape and appearance resemble those of the preceding, but they are larger and more convenient for use.

210. b. Vaccinium bracteatum, *Th.*, Jap. *Shashambo*, *Wakurawa;* an evergreen shrub of the order of Ericaceæ, occuring wild in warm regions. In autumn it produces many small dark purple berries in panicles, and they are subaciduous in taste.

210. c. Epigæa asiatica, *Max.*, Jap. *Iwanashi*, *Suna-ichigo;* a small evergreen shrub of the order Ericaceæ, principally found in the provinces of *Yamato*, *Yamashiro*, *Settsu*, and their vicinities. In spring it blooms small pink flowers in clusters and ripens large bean-sized fruits in summer. The fruits are covered with sand-like grains, and are soft, brittle, and subacid.

210. d. Empetrum nigrum, *L.*, Jap. *Gankōran;* a tiny evergreen shrub of the order Empetraceæ, growing wild on alpine regions. It has distinct male and female flowers on separate plants. Late in spring it blooms flowers, being succeeded with small purplish black subaciduous berries.

211. Elæagnus pungens, *Th.*, Jap *Natsu-gumi;* a

deciduous shrub of the order Elæagnaceæ, growing wild on mountains and in fields. It is also cultivated for its fruits. It attains a height of 8-9 fts., and late in spring it blooms flowers from the axils of the new leaves and bears red oval or sometimes round berries with white spots. The taste is subacid with a slight astringency.

212. Elæagnus umbellatus, *Th.*, Jap. *Aki-gumi*; a deciduous shrub of the order Elæagnaceæ, growing wild on mountains and in fields, attaining a height of about 10 fts. It yields fruits when still a young plant. Late in spring it blooms several flowers in clusters from the axils of the leaves, and is succeeded by red round berries with white starry spots. The taste is subacid with a slight astringency.

213. Elæagnus longipes, *A.* Gray., Jap. *Nawashiro-gumi*, *Tawara-gumi*; an evergreen shrub of the order Elæagnaceæ, attaining a height of about 10 fts. In winter it bears flowers from the axils of the leaves, and oblong red berries ripen in the beginning of the summer of the following year. The fruits are red and covered with micaceous starry spots, having a subacid and slightly astringent taste.

213. b. Cudrania javanensis, *Trecul.*, Jap. *Kwakwa-tsugayu*; an evergreen shrub of the order Urticaceæ, of a vine-like nature, provided with thorns on the stem, and found in the provinces of *Satsuma* and *Osumi*. The barren and fertile flowers shoot separately on distinct plants. It bears flowers in summer and reddish yellow sweet fruits in winter. They are eaten fresh or preserved in sugar. The wood is used for dying yellow.

213. c. Ribes grossularioides, *Max.*, Jap. *Suguri*; a small shrub of the order Saxifragaceæ, with slender stems attaining a height of 2-3 fts., provided with sharp thorns. The small flowers droop from the axils of the leaves, and the greenish aciduous berries are ripe in summer.

213. d. Ribes rubrum, *L.*, var. bracteosum, *Max.*, Jap.

Fusa-suguri; a small shrub of the order Saxifragaceæ, found in the forests of *Hokkaido*. It attains a height of 5-6 fts. The drooping flowers are disposed on panicles. The red berries have a subacid taste.

214. Ficus carica, *L.*, Fig, Jap. *Ichijiku;* a deciduous shrub of the order Urticaceæ, cultivated in warm countries, attaining a height of about 10 fts. In summer it produces fruits at the axils of the leaves. The flowers are concealed inside the fruits. The fruits are green and ovate at first, and then turn dark purple outside and reddish inside. It is soft, sweet, and slightly aciduous. Several good varieties have been recently introduced. Especially one with a greenish white skin is sweet like honey when ripe and is good for drying.

214. b. Ficus carica, *L.*, var., Jap. *Shiro-ichijiku, Nankin-ichijiku;* a variety of fig of a dwarf nature. The leaves and fruits are also small. When fully ripe the inside of the fruits is white and of an inferior taste. The fruits of Ficus pyrifolia and Ficus nipponica are also eaten.

215. Stauntonia hexaphylla, *Dec.*, Jap. *Mube, Tokiwa-akebi;* an evergreen climbing plant of the order Menispermaceæ, growing wild and also cultivated as an ornament and for it fruits. In summer it produces fine peduncles, and male and female flowers grow separately. The fruits ripen late in autumn. They are oval, and about 2½ inches long and 1 inch in diameter. They have a dark-red colour outside, and contain many black seeds. Their white pulpy flesh is sweet like honey. Formerly the fruits were presented to the Emperor as delicious and of great rarity brought from the province of *Omi*. They were highly prized at that time, as sugar was then unknown. The fruits of Akebia quinata and A. lobata are also eaten, and from these seeds oil is extracted.

216. Ginkgo biloba, *L.*, Jap. *Ginnan;* the nut of this plant is called *Ginnan*. This tree belongs to the order Coniferæ, having deciduous leaves and attaining a height of 40-50 fts. Barren and fertile flowers bloom separately on distinct plants. The leaves resemble the webbed foot of a duck. In spring flowers appear with the young leaves. The fruits ripen late in autumn. They are round and of a pale yellow colour. The nuts are obtained by taking off the fleshy substance. They are $\frac{1}{2}$-$\frac{2}{3}$ inch in length, and their kernels are eaten baked or boiled, or used in confectionery.

217. Torreya nucifera, *S. et Z.*, Jap. *Kaya, Kaye;* an evergreen tree of the order Coniferæ, growing wild every-where, but also cultivated for ornamental purposes. It attains a height of several feets. In winter it blooms male and female flowers separately on distinct plants. Its fruits ripen late in autumn. The fruits are oblong or oval with a resinous flesh covering the nut, which is first steeped in ash-water, then in fresh water, afterwards dried and preserved. The nuts consist of two sorts, round or oval. They are eaten raw or baked, and have an aromatic flavour. They are also used in confectionery or for taking oil.

217. b. Torreya nucifera, *S. et Z.*, var., Jap. *Shibunashi-gaya, Hadaka-gaya;* a variety of the former. The inner skin of the nut is attached to the shell, and the kernel is easily separated from it. This is called the bare Torreya nut, and is of the best quality. The provinces of *Mino, Iga,* and *Yamato* are noted for its production.

217. c. Pinus koraiensis, *S. et Z.*, Jap. *Chosen-matsu-no-mi;* the seeds of Pinus koraiensis (687). The kernels of these seeds are eaten and have a resinous aromatic flavour. The acorn is about 6 inches in length, and its seed about half an inch long.

217. d. Cycas revoluta, *Th.*, Jap. *Sotetsu-no-mi;* the seeds of this plant are produced among the leaves at the head of trunk. Several grains are attached to a brownish peduncle, and in autumn they ripen to a vermilion-red flat oval form about 1 inch long. The kernels have the taste of chest nuts and are eaten either fresh or dry.

218. Juglans sieboldiana, *Max.,* Jap. *Onigurumi, Ogurumi;* a deciduous tree of the order Juglandaceœ, attaining a height of 20-30 fts. In summer it produces male and female flowers, being succeeded with many fruits clustered together. The fruits resemble the peach in shape, and ripen in autumn to a black colour. The flesh is taken off, and the nuts are collected. Their form and size are different according to the species of the trees.

219. Juglans regia, *L.,* var. sinensis, *Casim.,* Jap. *Kuwashi-gurumi, Chosen-gurumi, To-gurumi;* this is closely allied to the preceding, but it has broader leaves and yields solitary fruits instead of clustered ones. When ripe the fruits burst themselves and expose the nuts which are large and round. Their shell is easily broken. They are delicious and used as a desert when dried.

220. Juglans cordiformis, *Max.,* Jap. *Hime-gurumi, Me-gurumi;* a species of Juglans. Its nuts are flat, narrow, and smooth with a shell. They are easier to break the shell than that of Juglans sieboldiana, and are much used as *Mukigurumi* (peeled kernels) for cooking and confectionery. They are also used for making oil, which is used for cooking and polishing wooden articles. The skin of the fruit is used for dying brown in the same way as Juglans sieboldiana.

221. Castanea vulgaris, *L.,* var. japonica, *D.C.,* Jap. *Kuri;* a deciduous tree of the order Amentaceae cultivated everywhere in the country, attaining a height of 30-40 fts. and sometimes 50-60 fts. It bears male and female flowers separately on the same tree in June. Its fruits ripen late in autumn. When they are ripen, they burst themselves and expose several nuts. The nuts are of different varieties. The kernels are eaten baked, steamed, or boiled, and they are also used in cookery and confectionery. Sometimes oil is extracted from them. The variety called *Tamba-guri* is the largest, and the one called *Hako-guri* contains several nuts in one fruit.

222. Castanea vulgaris, *Lam.*, Jap. *Shiba-guri, Sasa-guri;* the original species of chestnuts, growing wild. They are cut down every year, but they yield many fruits at the height of 1-2 fts. The one called *Yamaguri* (mountain-chestnut) grows to the height of 20-30 fts., and produces many fruits, which are used as food, and also dried and preserved. The fruits are about an inch in size.

223. Corylus heterophylla, *Fisch.*, Jap. *Hashibami, Kinchaku-hashibami;* a deciduous shrub of the order Amentaceae, growing wild and attaining a height of 6-7 fts., with barren and fertile flowers separately on the same plant. The flowers bloom early in spring, and the fruits ripen late in autumn. They are round and about ⅜ inch in size, with thick shells containing kernels, which are like chestnuts in taste.

224. Corylus rostrata, *Ait.*, var. sieboldiana, *Max.*, Jap. *Tsuno-hashibami, Naga-hashibami, Oni-hashibami;* a species of the preceding which resembles in shape, but the leaves are smaller. The acorns are concealed in a long slender covering. They are oval with a pointed head, and are used in the same way as the former.

225. Quercus cuspidata, *Th.*, Jap. *Shiinomi;* an evergreen tree of the order Amentaceae, growing abundantly in warm regions and attaining a height of 30-40 fts. It blooms in summer, and its fruits ripen late in the autumn of the following year. Its acorns are eaten parched. As the wood is hard and strong, it is used to make handles for oars.

226. Quercus glabra, *Th.*, Jap. *Matebashii, Satsuma-shii;* an evergreen tree of the order Amentaceae, found principally in warm regions, attaining a height of 20-30 fts. It blooms and bears fruits at the same time as the preceding. The fruit is oblong oval and about 1 inch in length. The acorn rests on a receptacle, and has the shape of that of the oak. It is delicious when parched.

226. b. Quercus gilba, *Bl.*, Jap. *Ichiigashi, Ichii;* an evergreen tree of the oder Amentaceae, growing principally in warm regions, attaining a height of 20–30 fts. The acorn is like that of Quercus acuta, *Th.* or *Q.* glauca, but is edible being less bitter.

227. Euriale ferox, *Salisb.*, Jap. *Onibasu;* an annual aquatic herb of the order Nymphaeaceae, growing in ponds and marshes, with broad round leaves floating on the water. It is green on the upper part and purple underneath, with thorns on both sides. In summer it bears flowers above the water surface, opening during the day and fading in the evening. After the flowers fade, the thorny balls grow to a size of 3–4 inches. They contain several round seeds about of the size of a finger. The seeds are collected for their edible kernels, which are dried and preserved and used to make starch. Its young stalks and roots are also edible.

228. Nelumbo nucifera, *Gaertn.*, Jap. *Hasu-no-mi;* the seeds of this plant are produced about 30 in number in a carpel. They are oblong and oval, being about of the size of a finger. The kernels are eaten fresh when they are green. The shell is black and hard when ripe, and has a white kernel inside. The kernel is dried and preserved by taking away the embryo which is bitter. It is used for cookery, and made into starch.

229. Trapa bispinosa, *Roxb.*, Jap. *Hishi-no-mi;* an annual aquatic herb of the order Onagraceæ, growing in ponds and marshes. Its leaves float on the water surface, and it blooms 4 petaled white flowers, being succeeded by 2–4 horned fruits, which are of different sizes and are used in the same way as the lotus.

230. Citrus nobilis, Mandarin orange, Jap. *Mikan;* an evergreen tree of the order Aurantiaceæ cultivated in warm regions, being about 10 fts. high. It blooms early in summer, and its fruits ripen in winter. The fruit-skin is of an orange colour, and incloses a juicy pulpy carpel. There are several varieties according to the climate in which they are cultivated. The district of *Yatsushiro* in the province of *Higo* is noted for the production of the frints, but the most celebrated place is the province of *Kii*. Some of the superior varieties of the orange family are as follows.

230. b. Citrus, Jap. *Yukō;* a variety of orange standing between Citrus nobilis, *Lour.*, and C. medica limonum, *Brandis.*, in form, taste, and flavour. So it is harder than the mandarin orange. It is good to eat, although somewhat inferior to the preceding.

230. c. Citrus, Jap. *Ujukitsu;* the shape of this fruit is round or pointed with a yellow skin. It is juicy, but not very sweet unless it is preserved till summer.

231. Citrus nobilis, *L.*, var., Jap. *Unshiu-mikan;* a variety of sweet orange with large fruits of about 3 inches in diameter and 1½ inches in height. It has a thin skin, few seeds, and a rich sweet juice. It is the best Japanese orange. Lately large quantities of these oranges have been produced in the province of *Kii*. There are several varieties of this sort. One called *Rifujin-kitsu* in *Kiushiu* and *Shikoku* belongs also to this species.

231. b. Citrus nobilis, *L.*, var., Jap. *Tōmikan;* a variety of the preceding with a thick warted skin and few seeds. It grows abundantly in the provinces of *Mikawa* and *Owari*, and also in other eastern parts. Though inferior in quality to the preceding, yet it is well fit for preservation.

232. Citrus aurantium, *L.*, var., sinense, *Galisco.*, Jap. *Kunenbo;* an evergreen tree of the order Aurantiaceæ, cultivated in warm provinces, being about 10 fts. high. It resembles the sweet orange in shape, but larger. In early summer it blooms fragrant white flowers, being succeeded by fruits which ripen in the following year. The fruit is about 2½ inches in diameter with a thick rind and nice flavour. Though not very sweet, yet it is preservative.

233. Citrus decumana, *L.*, var., Jap. *Jaga-tara-mikan;* an evergreen tree of the order Aurantiaceæ cultivvted in warm regions, being about 10 fts. high, with large leaves as those of the preceding. It blooms in summer and bears orange red fruits in winter. They are thick-skinned, being about 5 inches in diameter and 2½ inches in height, and are sweet and juicy.

234. Citrus nobilis, *L.*, var., Jap. *Kōji-mikan, Kōji;* an evergreen tree of the order Aurantiaceae cultivated in warm regions, being harder than the mandarin orange. It is about 10 fts. high, spreading over a space of more than ten steps, yielding many fruits. The fruits are smooth and thin skinned, and though moreacid than the mandarin orange yet they are noted for ripening earlier than others. When they are kept till March or April, they become very sweet. There are two varieties, one yellow and the other red.

234. b. Citrus nobilis' *L.*, var., Jap. *Beni-mikan;* the fruits of this species are round, flat, and beautiful with a smooth thin red skin. They are sweet and juicy. Those of the variety called *Beni-kōji* resemble them very much in appearance, but are larger and inferior in taste.

235. Citrus bigaradia, *Dupam.*, Bitter orange, Jap. *Daidai, Zadaidai;* an evergreen tree of the order Aurantiaceae, being about 15 fts. high. Its flowers bloom in summer and its yellow fruits ripen in winter. When left on the branches till the following year they turn green again; so they are called *Kwaiseitō* which means turning-green. The variety called *Kabusu* resembles much this. Both are round with a diameter of about 2½ inches. The Juice is pressed and used as vinegar, and is very strong. The young fruits are preserved in sugar, and is used in place of *Marubushukan* (241). The ripe fruits are also preserved in syrup. From the rind a fragrant oil called *Tōhiyu* (orange oil) is obtained.

235. b. Citrus aurantium, *Risso.*, var., Jap. *Amadaidai;* this resembles the *Kabusu* orange in shape and colour, but less aciduous, being esteemed for its juicy fruits. The *Toumikan* of *Tosa* and *Kinkunenbo* of *Satsuma* are the same varieties, but of a better quality. All these oranges are difficult to peel.

235. c. Citrus, Jap. *Natsu-daidai;* this is the product of the province of *Nagato*. The fruit is large, flat, and yellow. It has an aciduous juice, and is preserved for summer use.

235. d. Citrus, Jap. *Naruto-mikan;* this is the product of the province of *Awaji*. The fruit is round in form, and has rough yellow skin and rich aciduous juice. It is in the same quality and use as the preceding.

235. e. Citrus, Jap. *Tōdaidai;* this is found principally in southern and western provinces. The fruits are large, round, and pointed at the top, with an orange red skin, and they are very sweet.

236. Citrus japonica, *Th.*, Jap. *Kinkan, Marumikinkan* (kumquat orange); an evergreen shrub of the order Aurantiaceae cultivated in warm regions. Some of the largest specimens are 6-7 fts. high. Even when young it blooms in summer and yields round fruits which ripen in winter, having about the size of a finger, with yellow skin. Its pulp is sour, but the skin is sweet and fragrant. It is preserved in sugar. When the fruits are left on the branches till March or April of the following year they turn very sweet.

237. Citrus japonica, *Th.*, Jap. *Nagami-kinkan;* a variety of the kumquat orange with elliptical obovate fruits, which are used in the same way as the preceding.

238. Citrus decumana, *L.*, var., Shaddock or Pompelmos, Jap. *Uchimurasaki, Tōkunenbo, Buntan;* an evergreen tree of the order Aurantiaceae cultivated in warm regions, being about 10 fts. high. It blooms in summer and the fruits ripen in winter. The fruits are 6-7 inches in diameter, and 5-6 inches in height. They have thick skin and beautiful purplish pulp. They are the largest among the orange family, with an agreeable subaciduous taste, and they are eaten fresh.

239. Citrus decumana *L.,* Jap. *Zabon;* this is closely allied to the preceding, but its fruits are smaller, and the pulp is of a bluish white colour.

240. Citrus media, *Risso.,* var., chirocarpus, Jap. *Bushukan, Tebushukan;* an evergreen tree of the order Aurantiaceae cultivated only in warm regions. It blooms in summer, and yields fruits in winter. The fruits are yellow with several finger-like protuberances at the top. Their skin is very thick. They are highly odorant and may be preserved, but they are principally used for ornamental purposes.

241. Citrus media, *Risso.,* Jap. *Maru-bushukan;* a species with an oval form and pointed head, being about 6 inches in height. It has also thick skin with very little pulp; so it is not fit for deserts, but the thick skin is preserved in salt and eaten as vegetables. Its young fruits are preserved in sugar or syrup in the same way as the preceding.

242. Citrus, Jap. *Tachibana, Ukon-no-tachibana;* an evergreen tree of the order Aurantiaceae, being about 10 fts. high, with the fruits shaped like *Kōji* (mandarin orange, 234), but smaller and with thicker skin. The fruits are beautifully yellow with a slightly bitter and aciduous taste.

242. b. Citrus, Jap. *Sudachi, Riman;* this is produced in the provinces of *Kii, Awa,* and their vicinities, resembling the former in form, with thin skin and sour juice which is pressed out and used instead of vinegar. A variety produced in *Hizen* in the name of *Kinosu* resembles this very much.

243. Bromelia ananas, *L.,* Jap. *Ananasu;* an evergreen herb of the order Bromeliaceae produced in hot regions. It is cultivated in Loochoo and Bonin Islands. It must be kept in hot-houses in winter in temperate regions. The leaves are broad and flat, being 2-3 fts. in length. In summer it blooms among the leaves, and ripens yellow scaly oval fruits which are 5-6 inches in length. The fruits are sweet and of a nice flavour.

243. b. Musa sapientum, Jap. *Mibashō, Banana;* a species of Musa yielding edible fruits and purple flowers. It is cultivated in Loochoo and Bonin Islands. When the fruits are ripen, they are yellow and 4-5 inches in length. The pulpy flesh is edible by peeling off the skin, and sweet and fragrant. They are esteemed as the best fruits of the south. They are used to make alcohol and vinegar.

244. Jambusa vulgaris, *DC.*, Jap. *Hotō;* an evergreen tree of the order Myrtaceae, being about 10 fts. high, and produced in warm climates, as in Bonin Island. In temperate regions it must be kept in hot-houses during winter. In summer the flowers form a ball of numerous white stamens. The fruits resemble the loquat, with yellow skin and large seeds. They are very sweet and juicy.

244. b. Nephelium longan, *Lam.*, Jap. *Riu-gan;* an evergreen tree of the order Sapindaceae found in warm regions. It is cultivated in Loochoo Island. It is to be kept in hot-houses in winter. The leaves grow up pinnately on a petiole. The fruits are round and about the size of a finger. When ripen they are eaten either fresh or dried.

244. c. Nephelium litchi, *Camb.*, Jap. *Reishi;* very closely allied to the preceding, but the plant and fruits are twice in largeness. When ripen it has a beautiful red shrivelled skin. It is sweet and delicious. It is preserved longer than any other fruit. The plant does not thrive in cold places, but it grows in the southern part of *Osumi* province.

245. Citrullus edulis, *Spach.*, Water melon, Jap. *Suikwa;* an annual cultivated climber of the order Cucurbitaceae. It produces barren and fertile flowers separately on the same vine. Its fruits ripen in mid-summer. The fruit is larger than a man's head, with dark green skin and generally red pulp and black seeds. As the pulp contains plenty sweet liquid, it is eaten fresh, and when young it is preserved in salt and eaten as pickles. There are several sorts of colours and forms.

246. Citrullus edulis, *Spach.*, var., Jap. *Shiro-suikwu;* a variety of the preceding. The fruit has a white skin, yellow pulp, and red seeds.

247. Cucumis melo, *L.*, Melon, Jap. *Makuwa-uri;* an annual cultivated climber of the order Cucurbitaceae. It is produced much in the village *Makuwa* in the province *Mino*, whence the name is derived. The male and female flowers are separately on the same vine. The fruits ripen in mid-summer. They are oval-shaped, about 5 inches long, and of a yellow colour, with longitudinal stripes. They are eaten 1 or 2 days after having been collected, and are very sweet and delicious. There are several varieties of different colours and forms.

248. Cucumis melo, *L.*, var., Jap. *Ginmakuwa-uri, Ginmakuwa;* a variety of the melon with large fruits of a green rough skin. It is inferior in taste to the preceding.

249. Momordica charantia, *L.*, Jap. *Tsuru-reishi, Niga-uri;* an annual cultivated climber of the order Cucurbitaceae. The male and female flowers are separately on the same vine. The fruit ripens in summer. It is green, about 4 inches long and 2½ inches in diameter, and covered with irregular warts. It turns yellow when ripe, and bursts at the top and exposes several red pulpy seeds of the size of a finger. The plup is beautifully red, soft, and sweet, containing peculiar seeds. In the provinces of *Kiushiu* there is a variety with a long fruit about 2 fts. long called *Naga-reishi* (long *reishi*). The young fruits are eaten as vegetables.

Note.—The varieties of fruits here mentioned are only a few selected ones. Mume-plums, apricots, peaches, plums, and especially pears, persimmons, and oranges have a great many varieties which are too numerous to be mentioned respectively in this limited space. Moreover their cultivation increases new varieties continually. Besides these there are many plants yielding edible fruits. The following are the names of such plants;—Morus alba, *L.* (294), Pyrus toringo, *Sieb.*, var. incisa, *Fr. et Sav.* (362), Pyrus (363), Hovenia dulcis, *Th*, Cornus kousa, *Buerg.* (645), Prunus pseudocerasus, *Lindl.*, Cornus officinalis, *S. et Z.* (432), Taxus cuspidata, *S. et Z.* (580), Celtis sinensis, *Pers.* (558), Aphananthe aspera, *Pl.* (297), Opuntia ficus, *L.* (808), Ribes ambignum, *Max.* (809), Cornus canadensis, *L.* (824), Sterculia platanifolia, *L.* (599), Aesculus turbinata, *Bl.* (535), Quercus serrata, *Th.* (295), Q. glandulifera, *Bl.* (563), Q. acuta, *Th.* (564), Q. glauca, *Th.* forma sericea, etc.

CHAPTER X.—STARCH PLANTS.

This Chapter includes the plants which roots, stems, or seeds yield a white powder-like substance, which is made into starch. Starch is nutricious and used for making several sorts of food, as bread, paste, etc.

250. Apios fortunei, *Moench.*, Jap. *Hodo-imo;* a perennial climber of the order Leguminosae growing wild. The compound leaves have 3-5 leaflets on a common petiole. The vine is thin and is about 10 fts. long. In summer it produces greenish yellow papilionaceous flowers in panicles from the leaf-axils, being succeeded with pods about 2¼ inches long. The roots creep under ground with round bullet-like tubers. In winter they are dug out and eaten boiled. A kind of starch is also manufactured from them.

251. Pueraria thunbergiana, *Benth.*, Jap. *Kudsu*, *Makudsu;* a perennial climber of the order Leguminosae growing wild. The leaves are ternate, and the vine separates into many branches. In autumn it produces purplish red flowers in panicles, which are succeeded by flat hairy pods containing small seeds. The largest roots are 3-4 fts. and have about the thickness of a man's arm. In winter they are taken, and an excellent starch is prepared from them. It is used as food or paste. The vine is used to make baskets, and its fibre is taken for cloth. The leaves are used to feed cattle.

252. Trichosanthes cucumeroides, *Ser.*, Jap. *Karasu-uri, Tama-dzusa;* a perennial climbing plant of the order Cucurbitaceae growing wild every-where. The leaves are 3-5 lobed and hairy on the surface. The male and female flowers are on the different vines. In summer white flowers open, succeded by fruits of the size of a duck's egg. In winter they are taken and dried to be used for washing instead of soap. The seeds have a form as a clasped letter paper, whence derived the name *Tamadzusa* (letter). In winter the roots are collected to make starch.

253. Trichosanthes japonica, *Regel.*, Jap. *Kikarasu-uri, Gori;* a perennial climbing plant of the order Cucurbitaceae growing wild everywhere. The barren and fertile flowers open on the different plants. It resembles very much the preceding, but the leaves are lustrous on the surface and the fruits are twice as large when they ripe. The young fruits are eaten preserved in soy or salt. In winter the roots are collected and made into starch called *Tenkwa-fun*.

254. Dioscorea japonica, *Th.*, var. bulbifera, Jap. *Kashiu-imo, Ke-imo;* a perennial climbing plant of the order Dioscoreaceae growing wild or cultivated. The male and female flowers open separately on the distinct plants. The large round tubers on the branches are covered with warts. The roots are large, tuberous, and fibrous. They are used to prepare starch. As they are acrid, it is required to wash them with lye before eating.

255. Polygonatum canaliculatum, *Pursh.*, Jap. *Naruko-yuri;* a perennial herb of the order Smilaceae growing wild about 3 fts. high. The leaves are lanceolate with five longitudinal ribs, and from their axils are produced fine drooping peduncles with flowers, which are succeeded with round berries. In winter the roots are taken and made into starch. They are also eaten dried and preserved in sugar or syrup.

256. Polygonum vulgare, *Dest.*, Jap. *Amadokoro;* a perennial herb of the order Smilaceae growing wild about 2 fts. high. It resembles the preceding in shape, but harder. In winter the roots are taken to make starch.

257. Erythronium dens-canis, *L.*, Jap. *Katako-yuri, Kata-kuri;* a perennial herb of the order Liliaceae growing wild in cold regions. It has two leaves which are oval and pointed. A peduncle grows in the centre of the leaves and bears a flower like that of the lily. The roots are collected and made into starch which is coarsely grained and sticky. The starch is used to make vermicelli and cakes. The leaves and stalks are eaten boiled.

258. Orithia oxypetala, *Kunth.*, Jap. *Amana, Himesuisen, Tōrō-bana;* a perennial herb of the order Liliaceae growing wild. In spring it shoots two leaves, and in the midst of these a peduncle grows, which bears six petaled white flowers with dark purple veins. The roots are taken and made into starch, and the leaves are eaten as a vegetable.

259. Lilium cordifolium, *Th.*, Jap. *Uba-yuri, Gawa-yuri;* a perennial herb of the order Liliaceae growing on mountains. The small bulbs bear only 2-3 leaves without any flower, but the large bulbs have thick stalks about 2-3 fts. high with several leaves on the upper part, and produce 2-3 greenish white flowers at the top, facing laterally. The roots form scaly bulbs of the size of a large chestnut like that of the lily. They are collected to make starch for food. The young leaves are eaten boiled.

Note.—Besides these above mentioned, there are many plant which are rich in starch, as the root of ferns (91), batatas (109), pototo (109. b), Dioscorea (110), Colocasia (114), Sagittaria (118), Scirpus (120), lily (121), etc.; the roots of Nerine (509),

Lycoris (510), Alisma (514), et.; the stems of Cycas (710); the grains of rice (1), wheat (7), maize (19), Coix (20), buckwheat (46), etc.; the dry fruits of chestnuts (221—223), nelumbium (228), trapa (229), etc.

CHAPTER XI.—Forage Plants.

This Chapter includes the plants used for feeding cattle. The stalks of cereals and fabaceous plants and the leaves and roots of vegetables are good as fodder, but as these are of a limited quantity, various wild herbs are used for this purpose.

260. Medicago denticulata, Jap. *Uma-goyashi;* a biennial plant of the order Leguminosae growing wild in spring about 2 fts. long. The leaves are ternate, and from their axils fine peduncles are produced together with small yellow flowers which are succeeded with thorned and screw-shaped pods. It is highly relished by horses, whence the Japanese name is derived. It is not only used as a forage, but also eaten as a vegetable.

261. Medicago lupulina, *L.*, Jap. *Kometsubu-mago-yashi;* a species of the preceding of almost the same shape, but smaller und covered with hairs. As the seeds are like rice grains the name was derived. It is used in the same way as the former.

262. Melilotus suavoleus, *Ledeb.*, Jap. *Shinagawa-hagi, Yebira-hagi;* a triennial herbaceous plant of the order Leguminosae, growing wild in *Shinagawa* in the province of *Musashi*, whence it derives the name. It is about 3 fts. high, and in summer it produces small papilionaceous flowers, followed with small pods.

263. Vicia hirsuta, *Kock.*, Jap. *Sudzume-no-yendō;* a biennial herb growing wild. Its slender stems creep on the ground or climb to other things, being about 2 fts. long. In early summer it bears small white flowers on fine peduncles, being succeeded with small pods.

264. Vicia sativa, *Miq.*, var. angustifolia, Jap. *Karasu-no-yendō;* it resembles the former, but is larger. Its flower is purple, and pod larger.

265. Vicia tetrasperma, *Maench.*, Jap. *Kasuma-gusa;* a variety of No. 263, resembling in shape, with two light purple flowers on a fine peduncle.

266. Vicia cracca, *L.*, var. japonica, *Miq.*, Jap. *Kusa-fuji;* it resembles No. 263, but is larger. Its light purple flowers open in cluster.

267. Sonchus oleraceus, *L.*, Jap. *Nageshi, Keshi-azami;* a biennial herbaceous plant of the order Compositae growing wild everywhere, sprouting up at late autumn. During spring and summer of the following year it becomes about 2 ft. high, producing yellow composite flowers on branched stems. When the seeds are ripe, they are provided with papus and fly about in the air. The leaves and stems contain a milky juce which is bitter in taste. The young plants are eaten boiled.

268. Panicum viride, *L.*, Jap. *Yenokorogusa, Neko-jarashi;* an annual plant of the order Gramineae growing wild everywhere. It sprouts in spring, and in summer it becomes 1-2 fts. high. The flowers open in a panicle with long purplish and sometimes green brisky hairs like a fox-tail. In autumn the seeds ripen resembling much those of Panicum oplysmenus (16). They are eaten under the name of *Aoyagi.*

269. Panicum littoralis, *Sw.*, Jap. *Hiyegayeri;* a biennial graminous plant much found wild. It resembles Panicum oplysmenus (16), and attains a height of about 2 fts.

270. Panicum crusgalli, *L.*, Jap. *Inu-biye, No-biye;* an annual graminous plant growing wild everywhere, resembling Panicum oplysmenus (16) and having small seeds.

271. Panicum crusgalli, *L.*, var., Jap. *Keinu-biye. Midsu-biye, Kusa biye;* a different form of the preceding, growing in moist places. It has a stronger stem and a larger panicle,..

As both are only the wild forms of Panicum oplysmenus (16), they may grow either in moist or dry ground.

272. Eleusine indica, *Gaertn.*, Jap. *Ohi-shiba, Chikara-gusa;* an annual graminous plant growing wild in dry ground or at the road side, being a about a ft. high. The panicles are divided into several branches. The leaves and stem are tough and strong.

273. Panicum sanguinale, *L.*, Jap. *Mehi-shiba, Yatsu-mata-gusa;* an annual graminous plant growing wild everywhere, being 2–3 fts. high, with many branched panicles.

274. Avena fatua, *L.*, Jap. *Karasu-mugi, Chahiki-gusa;* a biennial graminous plant growing wild about 2 fts. high, blooming in panicles. The awn is large, and provided with long hairs twisted to the left. The grain is thin as wheat.

275. Bromus japonicas, *Th.*, Jap. *Natsu-no-karasu-mugi, Natsu-no-chahiki;* a biennial graminous plant frequently found wild, resembling the preceding, but more slender, being about 2 fts. high. The awns and hairs are smaller, and the seeds are also smaller, ripening later.

276. Glyceria caspia, *Trin.*, Jap. *Dojō-tsunagi;* an annual graminous plant growing wild. In early summer it is about 2 fts. high.

277. Arundinella anomala, *Steud.*, Jap. *Toda-shiba;* an annual graminous plant growing wild in clumps. In summer it attains to the height 2–3 fts. The panicle is divided into many fine branches and is about an inch long.

278. Trisetum cernuum, *Trin.*, Jap. *Kanitsuri-gusa;* a biennial graminous plant frequently occuring wild. In the beginning of summer it grows to the height of about 1 ft. with a panicle formed of many small divisions provided with fine long hairs. This is one of the earliest maturity of graminous plants.

279. Poa annua, *L.*, Jap. *Sudzume-no-katabira;* a biennial graminous plant much found at the road side in late autumn, attaining a height of 5-6 inches. It blooms in late spring, and its seeds ripen in early summer. This is one of the earliest maturity of graminous plants.

280. Poa fertilis, *Host.*, Jap. *Ichigo-tsunagi, Nirami-gusa ;* a biennial graminous plant growing with everywhere and attaining a height of 1-2 fts. in summer. It resembles very much the former in form, but is larger.

281. Paspalum thunbergii, *Kunth.*, Jap. *Sudsume-no-hiye ;* a biennial graminous plant much found wild, attaining a height of 1-2 fts. in summer. The panicles consist of several branches on a stalk.

Note.—Those mentioned in the above Chapter are only a few of the wild forage plants. There are great many other forage plants.

CHAPTER XII.—PLANTS FOR LUXURY.

This Chapter includes those plants which are next in importance to food-yielding plants and rather resemble spices and condiments in quality. They are indispensable for the luxury of mankind. Some of them are wholesome, but others not so.

282. Thea chinensis, *Sim.*, Tea, Jap. *Cha-no-ki ;* an everygreen shrub of the order Ternstaemiaceæ. Though it grows wild in mountains, it is extensively cultivated. It attains a height of 6-7 fts. in the wild state, but the cultivated plants are generally cut down and trained to a height of 2-3 fts. In late autumn the white flowers are produced, and the fruits ripen in the autumn of the following year. In early summer the young leaves are gathered for *Cha* (tea) which in prepared by steaming, rubbing, rolling, and drying up. It is prepared in many different ways, giving various teas as *Sen-cha, Matcha, Ryoku-cha,* (green tea), *Kō-cha,* (red or black tea), etc. Oil is pressed out from the seeds.

283. Thea chinensis, *Sim.*, var. macrophylla, Jap. *Tô-cha, Kikko-cha;* a species of the former with larger leaves and flowers. As its leaves are bitter more than those of the common tea, only the first shoots are used for making the common Japanese tea (*Sen-cha*), but they are well fitted to prepare the red-tea (*Kô-cha*).

284. Lycium chinense, *Mill.*, Jap. *Kuko;* a deciduous shrub of the order Solanaceæ. The stem is slender and flexible like a tendril, but when fully grown it becomes thick and about 10 fts. long. The leaves are narrow and soft, being about an inch in length. It blooms in spring, and the small red fruits ripen late in autumn. The stem is provided with thorns. A thorny variety is called *Oni-kuko* (devil lycium), and a little thorned one *Tô-kuko*, (Chinese lycium). The young leaves of both are used to make a kind of tea. They are also eaten boiled.

285. Acer tataricum, *Linn.*, var. ginnala, *Max.*, Jap. *Maira-cha, Kara-kogi;* a deciduous tree of the order Aceraceæ growing wild on mountains. It attains about 10 fts. high. It proudces male and female flowers separately on the same plant. Its young leaves are gathered and used for tea.

286. Cassia mimosoides, *L.*, Jap. *Kôbô-cha, Kitsune-no-binzasara, Nemu-cha, Ichinen-cha;* an annual herb of the order Leguminoseæ growing wild and also cultivated. Its stem attains about 2 fts. high. It bears small pinnate leaves and yellow flowers, followed with pods an inch long. The young stem and leaves are cut and dried as a substitute of tea. In shape it resembles *Kusa-nemu* (Aeschinome indica, *L.*) which is poisonous.

287. Akebia quinata, *Decne.*, Jap. *Akebi, Akebi-katsura;* a deciduous climber of the order Menispermaceæ growing wild. There are 5 leaves on a stalk, and male and female flowers on the same plant. A dark purple fruit ripens in autumn. It resembles very much the fruit of Stauntonia hexaphylla, but the former generally bursts when fully ripe. The young leaves are gathered, steamed, and dried up, and used as a substitute of tea. There is a species with three leaf-lets called *Mitsuba-akebi* (Akebia lobata, *Decne.*), having the same use.

288. Hydrangea thunbergii, *Sieb.*, Jap. *Ama-cha;* a half lignous shrub of the order Saxifragaceæ, growing wild or in gardens. From one root many stems grow in a group 3-4 fts. high, sprouting in spring and flowering in summer. The flowers are green at first, but turn red afterwards. The young leaves are gathered, steamed, rolled between hands, and dried up, and used to make a sweet beverage called *Amacha* (sweet tea). It is also mixed with Japanese soy to give a sweet taste.

289. Gynostemma cissoides, *Benth.* et *Hook.*, Jap. *Tsuru-amacha, Amacha-dsuru;* an annual climbing herbaceous plant of the order Cucurbitaceæ growing wild. Its trailing stem is slender and 5-7 fts long, with 5 leaves on a petiole. The leaves are used in the same way as the former.

290. Ligustrum japonicum, *Th.*, Jap. *Nedsumi-mochi;* an evergreen shrub of the order Oleaceæ growing wild 7-8 fts. high. In summer small white flowers appear, being disposed in a panicle at the tops of branches, and afterwards small dark purple globular berries are produced. The seeds are collected, roasted and used as a substitute of coffee. The seeds of Ligustrum ibota are also used in the same way. Besides these the seeds of Ilex latifolia and the roots of *Kiku-nigana* are used for the same purpose.

291. Saccharum officinarum, *L.*, Sugar cane, Jap. *Satō-kibi;* a perennial graminous plant cultivated in warm regions. It is 5-6 fts. high, with narrow leaves 2-3 fts. long. In hot regions it grows about 10 fts. high, with the stem more than 1 inch in diameter, and with rush-like flowers and seeds. In late autumn the stems are harvested, and their saccharine juice is pressed out to make sugar by refining. There are black, red, white, and other sugars, which are all used in a great quantity. The uncrystallized sugar or syrup is often used for the preparation of alcohol. Besides this there are sugar-maple, sugar-beet, sugar-sorghum, etc. as sugar-yielding plants.

292. Nicotiana tabacum, *L.*, Tobacco, Jap. *Tabako;* an annual herbaceous plant of the order Solanaceae cultivated in fields. Several varieties are produced in different places, and the ways of cultivation are different in every place. Generally the seeds are sown in spring and transplanted in fields in summer, but in warm regions they are sown in the beginning of winter and planted in fields early in the following year. In summer the stems are 4-5 fts. high and produce many flowers at the top. The stems are, however, generally cut at the top before flowering, and the leaves are taken off from time to time. The harvested leaves are dried and preserved to make smoking tobacco by cutting or rolling.

293. Humulus lupulus, *L.*, var. cordifolius, *Max.*, Hop, Jap. *Karahanasō;* a perennial climbing herbaceous plant of the order Urticaceae growing wild in mountainous districts of northern regions. It is much improved by cultivation. Male and female flowers open on separate plants. In summer the male plant produces flowers in loose drooping panicles, while the female plant grows scaly cones or catkins. At the base of the scale there are included small round seeds, which are bitter and fragrant, and constitute what are called hops used by brewers and bakers.

Note.—The processes of preparing tea, sugar, and tobacco are impossible to be described completely in these limited lines, and so they were briefly mentioned here. Besides those mentioned in the foregoing numbers, roasted barley, beans, and coix, and also cut and slightly roasted sea gardle are used in the same way as tea, and so these may be included in this chapter. The leaves of Sterculia platanifolia, pines, etc. are also used as a substitute of tobacco, but they are omitted here.

CHAPSER XIII.—ECONOMIC PLANTS OF DIFFERENT USES.

This Chapter includes useful plants for the mankind with different economic purposes, except those used as food.

294. Morus alba, *L.*, Mulberry tree, Jap. *Kuwa;* a deciduous tree of the order Urticaceae. The male and female flowers are produced on separate plants. It reaches to the height 20–30 fts. when growing wild, but when cultivated it is cut down to a certain height for the purpose to gather the leaves easily. In spring the flowers appear before the leaves. The leaves are of many different shapes, being produced from the two typical forms of entire and dentate edges. These leaves are necessary food for silk-worms. In summer the purplish red fruits ripen, and they are eatable with an agreeable subacid taste. The young leaves are used to make a kind of tea, and the bark-fibres are used for the preparation of paper.

295. Quercus serrata, *Th.*, Jap. *Kunugi;* a deciduous tree of the order Amentaceae growing wild on mountainous regions, but much cultivated for fuel. It grows about 10 fts. high. In early summer it produces male and female flowers separately on the same plant, and in autumn it produces acorns of the size of a thumb. The leaves are used to feed several worms producing silk. Other quercus species allied to this are used for the same purposes.

296. Equisetum hiemale, *L.*, Jap. *Tokusa;* an evergreen herb of the order Equisetaceae, growing wild or in gardens. The stalk is hollow and tabular with many joints, being abour 2 fts. high. As it is hard and rough, it is used for polishing various articles as wood and horn. The flowers are produced at the top of the stalk, resembling those of Equisetum arvense, *L.* The quality of the stalks for polishing purpose differs according to the place where produced. Those produced from the village *Waka-mori* of the district *Funai* in the province *Tamba* are most famous for their good quality.

297. Aphananthe aspera, *Pl,,* Jap. *Muku-no-ki, Muku-yenoki ;* a deciduous tree of the order Urticaceae growing wild every where. It attains a height of 20–30 fts. In spring the male and female flowers appear at the same time with leaves, and in autumn purplish black berries about in the size of a pea ripen. The berries are edible with a sweet taste. As the leaves are hard and rough, they are used to polish various articles as wood and horns. The leaves are preserved for this purpose by drying them during autumn. Those produced from the village *Terakuma* of the district *Funai* in the province *Tamba* are best for this purpose.

298. Juncus communis, *E. May.,* Jap. *I, Yu-gusa, Tōshin-gusa ;* a perennial herb of the order Juncaceæ. Though it grows wild, it is much cultivated in wet places. It is about 4 fts. high, yielding flowers and seeds at the top. It grows in groups, and in summer it is cut and dried. Its pith is used as the wick of the Japanese lamp. The stalks are also used for weaving mats.

299. Sorghum nigrum, *Bœm. et Schule.,* Jap. *Hōki-morokoshi, Hossu-morokoshi ;* the panicles of this graminous plant are used to make brooms and brush on account of their numerous long stiff branches. Their small grains are eaten.

300. Kochia scoparia, *Schrad.,* Jap. *Hahakigi, Hōki-gusa ;* an annual herb of the order Chenopodiaceæ cultivated in the field and garden. The stem is about 3 fts. high and is ramified into many branches. In summer it yields small apetalous flowers which are succeeded with small seeds. When full grown the stem is cut and dried and used as a broom. Its young leaves are edible when boiled, and its fruits are also consumed by the name of *Tompuri* in the province of *Ugo.* Besides this, the articles used for making brooms are Chamaerops hair, bamboo branches, straw and sorghum panicles, branches of *Yashio-*azalea and Lindera hypoleuca, stems of Pertya scandens, panicles of evergreen Eularia, fibrous roots of Ischaemum sieboldi, etc.

301. Sapindus mukurosi, *Gærtn.*, Jap. *Mukuroji,* *Tsubu;* a deciduous tree of the order Sapindaceæ cultivated in several countries, growing about 20 fts. high. It produces small flowers in panicles, which are succeeded with round fruits of the size of about ¾ inch. When fully ripen the outer covering with wrinkles is yellowish brown in colour and includes a round hard black seed. The extract of this covering or skin is used for washing, and the black hard seeds are used to make buddists' rosaries and playing buckles.

302. Gledistschia japonica, *Miq.*, Jap. *Saikachi;* a deciduous tree of the order Leguminoceœ. It grows wild or in gardens. It attains a height of 20–30 fts., the stem being provided with sharpe thorns. In summer it produces small flowers, which are succeeded with pods 9–10 inches long and 1 inch or more wide, containing small flat seeds. The juice of this pod is used for washing, and is said it cleans well without impairing the articles, and is much used to wash furnitures. The young leaves are eaten when boiled.

303. Ilex integra, *Th.*, Jap. *Mochi-no-ki;* an evergreen tree of the order Illicineæ. It grows wild, but is much cultivated in gardens. It attains a height of 20–30 fts. It bears male and female flowers on separate plants. In summer it opens yellowish white flowers which are succeeded with red pea-sized berries. Bird-lime is prepared from the bark by pounding.

304. Trochodendron aralioides, *S.* et *Z.*, Jap. *Yama-kuruma, Ō-mochi-no-ki;* an evergreen tree of the order Magnoliaceæ growing wild on mountains, attaining a height of about 10 fts. In summer it produces umbrella-like flowers which are succeeded with small pea-sized berries. From the bark of this tree bird-lime is prepared by pounding and washing several times.

305. Luffa petola, *Ser.*, Jap. *Hechima, Ito-uri;* an annual climbing herb of the order Cuturbitaceæ cultivated in fields. In summer it produces yellow flowers, male and female separately on the same plant. In autumn its fruits ripen, about

1½ fts. long and with a diameter of 3 or 4 inches. The inside of the pepo is filled with a fibrous web which is bleached till it becomes white and soft and is used as an washing article like a sponge. The fibres are also used for many other purposes as to line the inside of slippers and hats, and to make summer shirts. The young fruits and leaves are eaten as vegetables.

305. b. Luffa petola, *Ser.*, var., Jap. *Naga-hechima, Riukiu-hechima;* a variety of the preceding, with its pepo about 6 fts. long, and of the same use.

306. Lagenaria vulgaris, *Ser.*, Jap. *Hiôtan;* an annual climber of the order Cucurbitaceæ cultivated in fields. There are male and female flowers separately on the same plant. In summer evening it opens its white flowers, and closes them in the morning. Its fruits are ripen in autumn, and they are used to make liquid-vessels by taking out the soft pulp and seeds, after the pepoes were steeped in water, and drying afterwards. They resemble cucurbita pepoes (136), differing only in forms. The shape of the pepo is just like 2 balls with a narrow joint. The length of the fruit is about 1 ft. There are many varieties, and the common kind is eaten as the cucurbita pepo, but the variety bearing small fruits can not be eaten, having a bitter taste.

306. b. Lagenaria vulgaris, *Ser.*, var., Jap. *Ō-hiotan;* a variety of the preceding having a very large fruit. Generally only one fruit is left on each plant, and for getting a very large fruit several plants are grafted together.

306. c. Lagenaria vulgaris, *Ser.*, var., Jap. *Shaku-hiotan, Hisago, Tsuru-kubi;* a variety of 306 with a long neck at the top of the fruit just like a handle, its under part forming a round body. It is just like a dipper in form and is used as a dipper.

306. d. Lagenaria vulgaris, *Ser.*, var., Jap. *Hyaku-nari-hiotan;* a variety of 306 with small fruits about 4 inches long. It is used as a small vessel like the preceding. As the taste is bitter, it can not be eaten.

306. e. Lagenaria vulgaris, *Ser.*, var., Jap. *Sennari-hyotan;* a variety of the former, but with smaller fruits which are about 1½ inches long.

307. Gymnogongrus pinnulata, *Harvey*, Jap. *Tsuno-mata;* an algæ growing in group on rocks in water, attaining a length of 6-7 inches. It is forked in several parts. When fresh, the colour is purple, but when bleached it turns to a yellowish white. It is used as paste or to wash hair. It is eaten when boiled in a state of jelly. In the harbour *Chō-shi*, the jelly is called *Iinuma-konnyaku*. There is a variety with a very large size, a foot long and 3-4 inches broad, and a variety called *Ko-tsuno-mata* is about 2½ inches long growing in shallow water. They are all of a similar use.

307. b. Glœopeltis coliformis, *Harvey*, Jap. *Funori;* a species of algæ growing on rocks where the tide ebbs. Its form is like that of a hollow tube at first, but it is gradually divided into branches which are about 4 inches long. When fresh it is dark purple, but when washed and bleached it turns to a pale yellow colour. It is made into a flat sheet, and then it is called bleached *funori* and is used for its mucilaginous paste. It is also simply dried and eaten as food. The form is different according to the places where it grows. That found in the province of *Satsuma* is superior in form and quality.

307. c. Gymnogongrus (?), Jap. *Saimi, Hachijō-fu-nori;* a species of algæ resembling the preceding, but with solid stems. When bleached, it is pale yellow, hard, and strong in texture. It is used for its mucilaginous paste. It is produced abundantly in the Island of *Hachijō*.

307. d. Chondrus plotynus, *G. Ag.*, Jap. *Hotokeno-mimi;* a species of algae growing in the iceland of *Yeso* and the northern province of the main land. It resembles the large leaves of *Tsuno-mata* in form, though thinner with two divisions. When fresh it is dark purple, but it turns pale yellow when bleached. It is used for its mucilaginous paste.

CHAPTER XIV.—OIL AND WAX PLANTS.

This Chapter comprises the plants yielding oil, wax, lacquer, etc. The oil is used for food, lamps, and various industrial purposes. The wax is used to make candles and other articles. The lacquer is the necessary ingredient for lacquer work. Insect-wax is the production of insects.

308. Brassica chinensis, *L.*, var., Rape, Jap. *Abura-na;* a cultivated biennial plant of the order Cruciferœ. Late in autumn the young plants are produced, and late in spring they shoot up flower-stalks to the height 3–4 fts., sometimes 8–9 fts. Early in summer the ripen seeds are gathered and are called rape-seeds. An oil is extracted from the seeds, and it is used for cookery, lamps, and several other purposes. The flower-buds and leaves are eaten by boiling or preserving in salt.

309. Sesamum indicum, *L.*, Sesamum, Jap. *Gama;* an annual cultivated plant of the order Bignoniaceae. The seeds are sown late in spring, and the 4-sided stem grows 3–4 fts. high, bearing at leaf-axils labiate flowers which are succeeded with long capsules, splitting longitudinally when fully ripe. They contain a great many fine seeds. There are three varieties, black, white, and brown coloured. The latter variety is the best to take oil. The oil is principally used for dressing food. The grilled seeds are used to add to cakes, salads, etc.

310. Perilla ocymoides, *L.*, Jap. *Yegoma;* an annual cultivated plant of the order Labiatæ. Its seeds are sown late in spring, and its stems grow about 2 fts. high. It produces long panicles from its branches, bearing small white labiate flowers. In autumn the seeds ripen and are gathered to take oil. As this oil

has a drying nature, it is used to oil water-proof cloaks and
bamboo umbrellas, and also for cookery. It is often used to
mix with rape seed oil to prevent the freezing of the latter. The
seeds are used instead of Sesamum seeds on grilling and also for
feeding small birds.

311. Camellia japonica, *L.*, Jap. *Tsubaki, Yabu-tsubaki;* an evergreen tree of the order Ternstroemiaceae, growing in warm provinces to a height of 20–30 fts. Early in spring it produces red flowers which are succeeded by round fruits. The fruits ripen at the end of autumn, when the shell splits out and exposes 2 or 3 dark hard seeds, which are gathered for oil called *Tsubaki-abura or Kino-mi-abura*. The oil is used for food or industrial purposes. The Islands of *Idsu* produce a great deal of this oil. The seeds of Camellia sasanqua also yield an oil, which is called *Katashi-no-abura* in the provinces of *Kyushiu*, and is used for the smilar purposes. Tea-seeds give also an oil.

312. Ricinus communis, *L.*, Jap. *Tō-goma, Tō-no-goma;* an annual plant of the order Euphorbiaceae. The seeds are sown in spring and grow to a height of 8–9 fts. The leaves are broad and palmate. Male and female flowers are separated on the same plant. The fruits are of the size of a finger head and covered with small thorns. One fruit contains 3 seeds, which are oval and white and dark variegated. From the seeds a thick oil is pressed out, being used to put in ink for stamps and for medicine and industrial purposes.

313. Elæococca cordata, *R. Br.*, Jap. *Abura-giri, Dokuye, Korobi;* a deciduous tree of the order Euphorbiaceae, frequently cultivated in warm provinces. It grows about 10 fts. high, branching much. It is a diaecious plant. The leaves are large and broad, and 3, 5, or 7 lobed. The flowers appear at the head and are very pretty, having 5 pink petals. In autumn the ripe fruits are collected to make an oil. The fruits are round and contain 3–4 seeds in each. The oil is thick and poisonous, and is used for lighting and to make oil-paper.

314. Cephalotaxus drupacea, *S.* et *Z.*, Jap. *Inugaya, Hebo-gaya;* an evergreen tree of the order Coniferæ, growing wild everywhere in mountains. It is also cultivated for oil. The stem attains a height of about 20 fts. It is a diœcious plant. In April it blooms, and its fruits ripen late in autumn. The fruits are red, oval, and about 1 inch long. Oil is pressed out from the nuts, but it is only used for lighting, as it is poisonous.

315. Litsaea glauca, *Sieb.*, Jap. *Shiro-damo, Aka-damo, Shiro-tabu;* an evergreen tree of the order Lauraceae growing in warm regions. The stem attains a height of about 20–30 fts. The leaves are oval, tapering at both ends, and green on the upper, and white on the under side. In late autumn, small flowers appear on the branches. In the winter of the next year red bean-sized oval fruits are produced. From the kernels of the fruits an oil is pressed out. The oil is called *Tabu-no-abura* in *Kiushiu* provinces, and is used for lighting, but its quality is inferior.

316. Lindera praecox, *Blume.*, Jap. *Abura-chan, Muradachi;* a deciduous tree of the order Lauraceae, growing wild everywhere. Its stem attains to a height of about 10 fts. It bears yellowish white small flowers before its leaves appear in spring. Its fruits ripen late in autumn. The fruits are quite round and have the size of a small finger. Oil is taken from the kernels, and is used for lighting.

317. Lindera triloba, *Blume.*, Jap. *Ukonbana, Hataukon, Shiro-moji;* a deciduous tree of the order Lauraceæ, growing wild in mountainous regions of cold countries. The stem is about 10 fts. high. In spring it bears pale yellow small flowers before the leaves appear. It is a diœcious plant. The fruits ripen in autumn and are quite round, being about ½ inch in size. Oil is taken and used for the same purpose as the preceding.

317. b. Styrax japonicum, *S.* et *Z.*, Jap. *Yego, Chishano-ki;* the green fruits of this tree (546) ripen late in autumn. They have about the size of a bean, and enclose a dark yellow hard nut, from which oil is taken as the preceding.

318. Cinnamomum pedunculatum, *Nees.*, Jap. *Yabu-nikkei, Koga-no-ki, Kusu-tabu ;* an evergreen tree of the order Lauraceæ growing wild in warm regions. The stem is 20-30 fts. high. In summer long branched peduncles come forth from the axils of leaves, and produce yellowish flowers, being succeeded with black fruits late in autumn. From the kernels wax is taken for candles. The wax is oily and soft.

319. Ligustrum ibota, *Sieb.*, Jap. *Koba-no-ibota ;* a half-deciduous tree of the order Oleaceæ, growing wild everywhere. The stem is 5-6 fts. high. In summer small white flowers appear in panicles, and in winter purplish black fruits are produced. The kernels of this and other similar fruits are used for coffee. On the stems of this and other similar trees a waxy matter is accumulated by the action of insects. The wax prepared from it is hard and lustrous, and is used for various industries.

320. Rhus succedanea, *L.*, Jap. *Haji, Rō-no-ki, Haje-urushi ;* a deciduous tree of the order Anacardiaceæ cultivated in warm regions. The stem is about 10 fts. high. In summer small flowers appear in panicles on branches. In autumn the fruits ripen, which are round and flat and ⅓ inch in size. Wax is taken from them.

321. Rhus vernicifera, *D.C.*, Jap. *Urushi-no-ki ;* a deciduous tree of the order Anacardiaceæ, cultivated in cold regions, growing 20-30 fts. high. The leaves are large and compound, forming pinnates. In summer diœcious small flowers appear in panicles. The fruit is almost the same as the preceding, and also wax is taken. Lacquer juice is obtained from the stem by splitting it. The juice thus obtained is an important ingredient for lacquer wares.

322. Rhus trichocarpa, *Miq.*, Jap. *Yama-urushi ;* a variety of the former, growing wild everywhere. Its fruits are quite the same in shape and use as the former, though smaller.

322. b. Sapium sebiferum, *Roxb.*, Jap. *Tō-haje, Nan-kin-haje;* a deciduous tree of the order Euphorbiaceæ cultivated in warm regions, growing 20-30 fts. high. In summer it produces monœcious flowers, and late in autumn the fruits ripen. The fruit is ⅓ inch in size, and encloses 3 seeds. The fruit is covered with a white powder, which is used to make wax. A lighting oil is taken from the seeds.

Note.—There are still numerous plants giving oil. Some of the principals are Soy-bean (22), Ground nut (46), various species of Brassica, nuts of Torreya nucifera (217), Juglans (218, 219, 220), Hazels (223, 224), and Fagus sylvestris (516), and seeds of Cotton (327), Carthamus tinctorius (367), Sun-flower (854), &c., but these are not described here. Generally in Japan Cereals and Legumes are not used for oil, but in China Soy-bean is much used for this purpose. Various oils used as medicine are also omitted here.

VOLUME II.

CHAPTER XV.—TEXTILE PLANTS.

This Chapter contains the plants, which give fibrous and flexible materials for threads and clothes. Therein are also included those, which stems, barks, and leaves being flexible are used for ropes, nets, brush-hairs, &c.

323. Cannabis sativa, *L.*, Hemp, Jap. *Asa ;* an annual plant of the order Urticaceae. Its seeds are sown in spring, and when fully grown its four sided stem is 7-8 fts. high. It is diaecious. In late summer the stems are harvested, and the bark is peeled off for thread and cords, and also for paper. The peeled stems are reduced to charcoal and used for gunpowder. The seeds are used as a spice or to make oil.

324. Baehmeria nivea, *Hook.* et *Arn.*, China grass, Jap. *Karamushi, Mao ;* a perennial herb of the order Urticaceae. It is growing wild everywhere, but it is better to be cultivated in fields. In spring its stem grows 3-4 fts. high, having the two sexes of flowers separately on the same plant. Late in summer the stem is cut off, and its bark is peeled for fibre. In warm regions it is cut 3 times and in cold parts twice in one year. There are several varieties, from all of which fibre may be taken.

325. Urtica thunbergiana, *S.* et *Z.*, Jap. *Irakusa ;* a perennial Urticaceous plant growing wild. In spring it shoots forth a stem to a height of 2-3 fts. Its leaves and stems are provided with thorns which sting sharply like a wasp. Its leaves are disposed alternately or opposit on the stem. In late summer the stems are cut for fibre, which is used for the same purpose as the preceding. Those growing in mountains and in *Hokkaido* have a height of 6-7 fts. The soft young stems and leaves are eaten by boiling.

326. Ulmus montana, *Sm.*, Jap. *Atsushi, Atsuni, Ohiyodamo ;* a deciduous tree of the order Urticaceae growing wild in mountains of northern provinces. The stem is about 10 fts. high. Early in summer, it bears bunches of small green flowers, which are succeeded with small flat scale-like fruits. The stiff bark of the stem is peeled off and used for fastening instead of a rope. In *Yezo* the fibre is used for weaving

327. Gossypium indicum, *Lam.*, Cotton, Jap. *Kiwata, Wata ;* an annual cultivated plant of the order Malvaceae. It is sown in spring, and attains to a height of 2-3 fts. in summer. A yellow flower is produced in each eaf-axil, and it is succeeded with a fruit in the form of a peach. When the fruit ripen the capsule bursts and cotton is exposed. The quality of the fibre is different according to the varieties, but it is all used for spinning. The oil pressed out from the cotton seeds is used for cookery and lighting.

328. Abutilon avicennae, *Gærtn.*, Jap. *Ichibi, Kiriasa ;* an annual cultivated plant of the order Malvaceae. It is sown in spring and attains to a height of 4-5 fts. In summer, it bears 5 petaled small yellow flowers in each axils of leaves, being succeeded with fruits. The fibre got from the bark of the stem is white and silky. The peeled stem is burnt to charcoal and used as a tinder.

329. Hibiscus syriacus, *L.*, Jap. *Mukuge, Hachisu ;* a deciduous shrub of the order Malvaceae cultivated in fields, growing 6-7 fts. high and in group. It is a good plant for hedges, and thrives well in wet places. Late in summer, its flowers open in the morning and fade in the evening. They are of different colours and of single or double petals. The bark gives fibre, which is also used to make *Mino* (farmers' rain coat), which is a famous production in the provinces of *Hōki* and *Inaba*.

330. **Hibiscus hamabo**, *S.* et *Z.*, Jap. *Hamabō;* a deciduous plant of the order Malvaceae growing wild along the sea coasts in warm provinces. Its stem attains to a height of about 10 fts. Early in summer, it blooms at the top of the branches and in the axils of the leaves. The flowers are like those of cotton, with yellow petals, purple at the base. A strong fibre is got from the bark, and is used as a rope.

331. **Urena sinuata**, *L.*, Jap. *Bondenkwa;* a deciduous shrub of the order Malvaceae, growing in warm regions and attaining to a height of 2-3 fts. The leaves are cut into 5 segments with green and white variegation. In autumn it blooms bright crimson flowers. The fibre taken from the bark is used as a rope.

332. **Corchorus capsularis**, *L.*, Jute, Jap. *Tsunaso*, *Ichibi;* an annual cultivated plant of the order Tiliaceae. It is sown in spring and grows to a height of 3-4 fts. In summer it yields small yellow flowers, which are succeeded with fruits. A strong coarse yellowish gray fibre is got from the bark, and is used as thread. The texture woven from this fibre is used for matting or wrapping proposes.

333. **Tilia cordata**, *Mill.*, var. japonica, *Max.*, Jap. *Shina-no-ki*, *Mada-no-ki*, *Hera-no-ki;* a deciduous tree of the order Tiliaceae growing wild in mountains 20-30 fts. high. In early summer it bears yellowish white fragrant flowers, which are succeeded with small round seeds. The strong bark is used as a rope, and the fibre of the young plant is woven into cloth. It seems that the fibre was much used in ancient times. There are several varieties, and that with large leaves grows quickly.

333. b. **Acer rufinerve**, *S.* et *Z.*, Jap. *Urihada-kayede;* a deciduous tree of the order Aceraceae growing wild in mountains of northern regions. The fibre of its bark is used for the same purpose as the preceding.

334. **Wisteria chinensis**, *S.* et *Z.*, *S.* et *Z.*, Jap. *Fuji*, *Yama-fuji;* a deciduous climbing plant of

the order Leguminosae, growing wild in mountains and also planted in gardens. In late spring it produces elegant purple or white papilionaceous flowers drooping in a raceme with the leaves. Afterwards long pods are produced. The branch is strong and flexible, and is used for fastening. The fibre taken from the bark is used for thread or cloth. The young tendrils when bleached are used for making baskets, etc. The young leaves and flowers can be eaten as vegetables. The seeds are also eaten when grilled. *Noda-fuji* (611) has the same uses.

334. b. Pueraria thunbergiana, *Benth.*, Jap. *Kudsu*; the fibre of the bast of this climber (251) is strong and white, and is used for weaving, and also to make ropes and nets. The twine is used to make baskets.

335. Cocculus thunbergii, *Dc.*, Jap. *Tsudsura-fuji*; a deciduous climber of the order Menispermaceae growing wild in bushes. The two sexes of flowers are produced separately on different plants. Late in spring, it produces yellowish green flowers, which are succeeded with round pea-sized black berries. The thin tendrils are bleached and used to make baskets, etc. which are the famous products of *Midsuguchi* in *Ōmi*.

335. b. Akebia lobata, *Dec.*, Jap. *Mitsuba-akebi*; a deciduous climber of the order Lardizabalaceae growing wild in mountains. The leaves are triphyllous, and the flowers opening in early summer are monaecious and are succeeded with edible fruits. In *Midsuguchi* of *Ōmi* and *Tsugaru* of *Mutsu*, the young vines are bleached and used to make baskets, etc. as the former.

336. Marsdenia tomentosa, *Morr.* et *Decne.*, Jap. *Fuyō-ran*, *Kijō-ran*; an everygreen climbing plant of the order Asclepiadaceae growing wild in mountains of warm regions. The leaves are round and smooth pointed at the apex. The flowers bloom in the axils of the leaves. The fruit forms a long follicle, which discloses a tuff of silky fibres. As the vines are strong and tenacious, they are used for making ropes and bow-strings.

337. Musa basjoo, *Sieb.*, Jap. *Bashō;* a perennial herbaceous plant of the order Musaceae cultivated in gardens. Late in spring it shoots forth its leaves to a height 10 fts. In summer it bears yellow flowers protected with large bracts. They are succeeded with fruits, which, however, do not come to maturity unless in hot climates. During the frosty season the leaves wither and only the sheath remains; so it must be covered during the severe winter. From the sheath, fibre is obtained. In the *Okinawa* Islands, a different species of Musa grows plentifully, and from its fibre, the natives weave a cloth called *Basho-fu* (Musa linen).

338. Juncus balticus, *Denth.*, Jap. *Kohige;* a perennial herb of the order Juncaceae cultivated in paddy fields. It grows to a height of about 3 fts. In summer its stalk bears small flowers arranged in branches under about 3 inches from the top of the leaves. The stalks are cut and woven into mats called *Bingo-omote*. They are also used to make hand-baskets. The Juncus (298) is of the same use, but softer.

339. Typha angustifolia, *L.*, Jap. *Himegama;* a perennial aquatic plant of the order Araceae growing in swamps and ponds. In spring it shoots forth broad flat leaves to a height of 6-7 fts., and in summer the flower-stalks grow to the height of the leaves, bearing male and female flowers separately at the top. The latter is in the form of a spadix of a length of 8-10 inches and a diameter of an inch. When fully ripen, the flower flies off by the wind like cotten. It is used for tinder, candle-wick, paper-making, etc. The leaves are used for mats, baskets, and ropes. There are two varieties, large and small, which are of the same use. The young leaves can be eaten as a vegetable.

340. Cyperus nutans, *Vahl.*, Jap. *Shichido, Riukiu;* a perennial aquatic plant of the order Cyperaceae cultivated in paddy fields. In spring, triangular stalks grow to a height of 4-5 fts., bearing fine flowers arranged on little petioles. In autumn they are cut, dried and woven into green mats called *Riukiu-omote* or *Ao-mushiro*.

341. Carex dispalatha, *Boott.*, Jap. *Kasa-suge;* a perennial aquatic plant of the order Cyperaceae cultivated in paddy fields. The leaves are flat, about ½ inch broad, and 3 fts. long. In summer it shoots forth stalks, which bear male and female flowers. In autumn the leaves are cut and made into a kind of hats which is a famous product of the provinces of *Kaga* and *Etsū*.

342. Carex, Jap. *Mino-suge;* a perennial plant of the order Cyperaceae growing wild in swamps. It resembles the former in form, but the leaves are more narrow and strong, being used to make farmer's rain coats. Besides these, there are many growing wild in mountains, and they are used for the same purpose.

343. Carex pierotii, *Miq.*, Jap. *Shio-kugu, Hama-kugu;* a perennial herb of the order Cyperaceæ growing wild in salty marshes. The length of the leaves are about 2 fts. In summer it shoots forth stalks bearing flowers at the head. In autumn the leaves are cut, dried and used to make ropes, etc.

344. Carex, Jap. *Yama-kugu;* a perennial herb of the order Cyperaceæ growing wild. Its form and use are as the preceding.

345. Elymus arenarius, *L.*, Jap. *Tenki, Hama-dō, Kusa-dō;* a perennial grass of the order Gramineæ growing wild on the sea coasts of northern regions. The leaves are about 3 fts. long and ½-⅔ inch broad, and covered with white powder. In summer it bears panicles of flowers. In the province of *Ugo* it is cultivated and used for making mats. The natives of *Yeso* use it for weaving, being called *Tenki*. The leaves being long are used for weaving and for making ropes and paper.

346. Hydropyrum latifolium, *Grisb.*, Jap. *Makomo;* a perennial grass of the order Gramineæ growing in moist ground. In summer the sprouts resembling those of bamboo grow to a height of 5-6 fts., and in autumn long stalks with male and female flowers at the top are produced. The seeds are used as food, and the new sprouts are also edible. A kind of mats is made from the leaves.

347. Imperata arundinacea, *Cyrill.*, Jap. *Chigaya;* a perennial wild grass. In late spring it produces flowers in panicles, which when young are eaten by children with the name of *Tsubana* or *Chibana*, and when fully ripen their soft fires are used instead of cotton or as tinder. From the leaves a mat and a farmer's rain-coat are made.

348. Andropogon schoenanthus, *L.*, Jap. *Ogaru-kaya;* a perennial wild grass. In autumn its stalk bears flowers, with awns twisted like oat. The fibrous roots are white and strong, and used to make brushes, etc. There is a variety called *Megaru-kaya*.

348. b. Iris ensata, *Th.*, var. chinensis, *Max.*, Jap. *Neji-ayame;* the fine fibrous roots of the Iris (933) are used in the same way as the preceding.

349. Bambusa aurea, *Sieb.*, Jap. *Usen-chiku*, *Hōrai-chiku;* a small bamboo, growing to a height of about 10 fts., and often used for hedges. In summer the young sprouts are eaten. As the stem is tenacious, it is used instead of ropes.

349. b. Bambusa, Jap. *Take-no-kawa;* the sheathes protecting the young bamboo sprouts, expecially those of *Madake* (589) and *Hachiku* (592) are much used for wrapping articles or for making Japanese slippers. Those of *Shiratake* (white bamboo) in the province of *Chikugo* are the best for slippers, as they have no spot.

349. c. Chamaerops excelsa, *Thumb.*, Jap. *Shuro;* the leaves of this palm (711) are used for plaiting purposes or for making brooms. Its stalk may be used for other purposes. Its hairy fibres are strong and water proving, and are much used for making ropes, mats, brushes, and many other articles.

Note.—Besides those above mentioned, the stems, straws, and vines of many plants have fibres: for Example, the straw of rice, barley, wheat, etc.; the leaves of Scirpus (964), Eularia (972), Amomum (128), and Ananas (143); the roots of Osmunda (91) and Sagittaria (118); the petioles of Lotus (125); the vines of melons; the bast of Lespedeza (612), Sophora (414), Sterculia (599), and Salix (658).

CHAPTER XVI.—PAPER PLANTS.

This Chapter comprises the plants giving raw materials and pasty fluids for making paper. Generally any kind of fibre may be used for making paper, but here are concerned only those commonly employed.

350. Broussonetia papyrifera, *Vent.*, Jap. *Kōzo, Kaminoki;* a deciduous tree of the order Moreæ cultivated extensively. It is cut off every year, so that it is only 6-7 fts. high. It is a diœcious plant, and the female flowers produce round fruits. In winter the stems are cut down and the barks are stripped off as an important material for paper.

351. Broussonetia kajinoki, *Sieb.*, Jap. *Kajinoki;* a deciduous wild tree of the order Moreæ growing 20-30 fts. high. It is a diœcious plant, having the same form as the preceding. The use is also the same, though inferior. The ripe fruits are beautifully red and sweet.

351. b. Morus alba, *L.*, var. stylosa, *Bur.*, Jap. *Kuwa;* this plant (294) being the same genus with the paper mulberry, a good paper may be manufactured from the bast, but as this plant is used especially for feeding silkworms, the paper made from the branches after the leaves are taken off for silk-worms is of a very inferior quality.

352. **Broussonetia kæmpferi**, *Sieb.*, Jap. *Mukumi-kadsura, Tsurukago ;* a deciduous climber of the order Urticaceæ growing wild in warm regions. It is a species of 350 of a climbing nature. Paper is manufactured with the fibre of the bark.

353. **Edgeworthia papyrifera**, *S.* et *Z.*, Jap. *Mitsumata, musubiki ;* a deciduous shrub of the order Thymeleaceæ cultivated in many countries. The stem is about 7 fts. high, and its branches are divided into three parts. Late in autumn, after the fall of the leaves, buds come forth in tufts at the head of each branch. In spring yellow flowers open, and then leaves come out. The branches are cut in autumn, and the bark is steeped in water, cleaned from the coarse part, and used for paper making.

354. **Wikstroemia canescens**, *Meisn.*, var. pauciflora, *Fr. et Sav.*, Jap. *Ganpi ;* a deciduous shrub of the order Thymeleaceæ growing wild in warm countries, of 5-6 fts. high. In summer, it produces flowers with yellow limbs and white tubes. The plants are pulled out during spring and autumn, and the bark is taken for paper.

355. **Wikstroemia japonica**, *Miq.*, Jap. *Kiko-ganpi, Hinoo ;* a deciduous shrub of the order Thymeleaceæ produced in warm regions, growing 3-4 fts. high. It resembles the preceding in form, but smaller and with yellow flowers. It is also used in the same way.

356. **Wikstroemia canescens**, *Meisn.*, var. *Ganpi.*, Jap. *Ko-ganpi ;* a low deciduous shrub of the order Thymeleaceæ growing wild 1-2 fts. high. It resembles the former in shape, but smaller, with white flowers shaded with pink. It is used for the same purpose.

356. b. **Daphne pseudo-mezereum**, *A. Gray*, Jap. *Oni-shibari, Sakura-ganpi ;* the bark of this plant (502) has fibre of a superior quality and is used for making paper.

357. **Hibiscus manihot**, *L.*, Jap. *Tororo-aoi, Neri ;* an annual plant of the order Malvaceæ. It is sown in spring, and

grows to a height of 3–4 fts., producing 5 petaled yellow flowers. That commonly cultivated is a dwarf variety, being about 1 ft. high, but with big roots. In summer the roots are dug out, dried, and used as a paste for manufacturing paper.

358. Hydrangea paniculata, *Sieb.*, var. minor, *Max.*, Jap. *Norinoki, Kineri, Nibenoki ;* a deciduous shrub of the order Saxifragaceæ growing wild in mountains. It attains to a height of 7–10 fts. In summer it bears flowers in clusters at the top of the branches. The bast of the stem and branches is used directly or after drying for pasting paper

359. Acer cratægifolium, *S. et Z.*, *Uri-kayede, Myō-ri-no-ki ;* a deciduous shrub of the order Aceraceæ growing wild in mountains, attaining to a height of 10 fts. In spring it blooms and sproats at the same time. It is a monæcious or diœcious plant. The bark of this tree is used in *Suruga* province as a paste in paper-making in summer by steeping it in water.

359. b. Sterculia platanifolia, *L.*, Jap. *Ao-giri ;* as the bark of this tree (599) gives white and strong fibre, it is used to weave cloth and to make ropes, but it is also used in paper-making on account of its rich content in a mucilaginous fluid.

359. c. Kadsura japonica, *L.*, Jap. *Binan-kadura ;* as the vine of this plant (403) is rich in a mucilaginous juice, it was only used for hair-dressing, but in *Satsuma* province it is used in pasting an inferior paper.

Note.—The vegetable fibres used for paper-making are not confined to those mentioned here. All the plants having fibrous barks, i.e. those contained in the division of textile plants might be used for this purpose. Especially the straw of rice, wheat, and other grasses, as well as coniferous timbers are lately used. There are still other plants giving pastes.

CHAPTER XVII.—Dye Plants.

This chapter includes the plants giving various dyes from their flowers, fruits, leaves, stems, or barks. Here are only mentioned those which are cultivated or grown wild and much used.

360. Berberis chinensis, *Desf.*, Jap. *Megi;* a deciduos shrub of the order Berberidaceæ growing wild in mountains 5-6 fts. high. The branches grow very thickly and are provided with fine thorns. In spring leaves and then flowers are produced, being succeeded with red berries which turn black when fully ripe. The bark of the stem is used for dying.

360. b. Evodia glauca, *Miq.*, Jap. *Kiwada;* the deep yellow bark of this plant (538) is used as a dye and medicine.

361. Isatis japonica, *Miq.*, Jap. *Hatokusa;* a biennial herbaceous plant of the order Cruciferæ introduced from China at the age of Kyoho. The leaves resemble those of rape, and are covered with white powder. Late in spring, yellow flowers open on stalks, and flat pods are produced. The leaves are used as a green dye.

362. Pyrus toringo, *Sieb.*, var. incisa, *Fr.* et *Sav.*, Jap. *Yama-nashi, Dsumi, Ko-nashi;* a deciduous wild tree of the order Rosaceæ, growing 5-6 fts. and sometimes 20-30 fts. high. Late in spring it bears pink flowers, which are succeeded with small round red or yellow berries of a strong acid taste. The bark is used as a yellow dye.

363. Pyrus, Jap. *Ōdsumi, Su-nashi, Kata-nashi;* a deciduous tree of the order Rosaceæ growing wild in mountains 20-30 fts. high. In the beginning of summer it produces 5 petaled white flowers shaded with pink. They are succeeded with round and about one inch sized berries which fall off in autumn. The berries are red and very aciduous, but they can be eaten by boiling or preserved by drying, being called *Sanzashi*. The dried thick bark of the stem gives a yellow dye called *Dsumi* by boiling it with water and coagulating the extract with alum.

364. Rubia cordifolia, *L.,* Jap. *Akane;* a perennial wild climber of the order Rubiaceæ. The petioles and tendrils are much provided with recurved prickles. In summer it produces small white flowers, which are succeeded with small round black berries. Formerly the roots were collected in winter and used for red dying.

365. Galium verum, *L.,* Jap. *Kawara-matsuba;* a perennial wild herb of the order Rubiaceæ, growing about 1.5 fts. high. In autumn it produces small yellowish white flowers disposed in panicles at the top of the young branches. In winter the roots are collected and used as a red dye. The roots sold as madder in commerce are mostly those of this plant.

366. Gardenia florida, *L.,* Jap. *Kuchinashi;* an evergreen shrub of the order Rubiaceæ. It is grown wild in warm regions, but much planted in gardens. The stem is 6-7 fts. high. In summer it bears 6 petaled white flowers, which turn yellow afterwards. The fruits are oblong and tapering at both ends, with longitudinal angles. The fully ripen deep yellow fruits are preserved after drying and used as a yellow dye or medicine. This tree is esteemed for gardens on account of the lustrous leaves and fragrant flowers. The petals are eaten as a vegetable.

367. Carthamus tinctorius, *L.,* Safflower, Jap. *Beni-bana, Suye-tsumu-hana;* a biennial cultivated plant of the order Compositae, growing in summer 4-5 fts. high. The stems and leaves are provided with sharp thorns. The reddish yellow flowers produced at the head of the branches are collected early in every morning and dried to make a red dye called *Beni,* which is used as a cosmetic by women. The seeds give an oil, and the young leaves serve as a vegetable.

368. Dyospyros lotus, *L.,* Jap. *Shina-no-ki, Saru-gaki, Mame-gaki;* a deciduous tree of the order Ebenaceæ cultivated in cold regions. The stem is 10 fts. or more high. After new leaves perfect or diaecious flowers are produced. The fruits are round or oblong, being about one inch long. The unripe fruits are collected and pressed to get *Shibu*, an astringent juice, which is used to give a brown colour to paper, cloth, wood, etc. by painting it on them and to protect them from rottening. The fully ripe fruits are edible. The centre black wood is called *Kuro-kaki* (black persimmon). The *Shibu* is obtained also from other astringent persimmons.

369. Ilex pedunculosa, *Miq.*, Jap. *Soyogo, Suzukashi;* an evergreen wild tree of the order Aquifoliaceæ, growing about 10 fts. high. It is a diaecious plant. The male flowers are small, white, and in clusters, and the female flowers are loosely arranged and produce small round red fruits. A brown dye is got by boiling the leaves.

370. Lithospermum erythrorhizon, *S.* et *Z.*, Jap. *Murasaki, Nemurasaki;* a perennial herb of the order Boraginaceæ, grown wild or planted in gardens, attaining to a height of about 2 fts. The head of the branches bears small white flowers, which produce small round seeds. In winter the roots are collected and dried for a purple dye. The wild roots are superior to the planted.

371. Basella rubra, *L*, Jap. *Tsuru-murasaki;* an annual cultivated climber of the order Chenopodiaceæ. The leaf-axils produce branches, which bear loosely arranged flowers and then pea-sized small round deep purple berries. The purple dye got from the berries is very fine, but it is liable to fade. The leaves are edible as a vegetable.

372. Polygonum tinctorium, *L.*, Jap. *Ai, Tadeai, Aitade;* an annual herbaceous plant of the order Polygonaceæ commonly cultivated in dry fields and sometimes in paddy fields, growing 1-2 fts. high. It produces pink flowers disposed in spikes, and then seeds. The leave are long and narrow, oval, etc. according to the different varieties. The dried leaves are made into indigo balls.

373. Mercurialis leiocarpa, *S.* et *Z.*, Jap. *Yama-ai;* an evergreen herbaceous plant of the order Euphorbiaceæ growing wild in shady places 1–2 fts. high. In summer it bears yellowish green flowers of the two sexes on the separate plant or on the same plant. In former times the juice of this plant was used to print on clothes in blue colour, but this plant does not contain enough the colouring matter to make indigo balls.

373. b. Justicia, Jap. *Riukiu-ai;* an evergreen herb of the order Acanthaceæ produced in *Riukiu* islands. Several stems grow in group from one root, attaining to a height of 1 or 2 fts. The leaves are oblong oval and of a bright dark green colour. It bears flowers very rarely. In *Riukiu* it is cut several times in a year and made into indigo. As this plant contains a great quantity of good indigo, it is now cultivated also in the southern countries.

374. Myrica rubra, *S.* et *Z.*, Jap. *Yama-momo;* an evergreen tree of the order Amentaceæ growing wild in warm regions, attaining to a height of 10–20 fts. It is a diœcious plant. In spring, the male flowers appear in the form of aments, and the female in short spikes. The fruits are of a purplish red colour with an agreeable sweet taste. There is a variety which produces white fruits. The bark which has the name of *Momo-kawa* or *Shibuki* contains much tannin and is used for a brown dye.

374. b. Machilus thunbergii, *S.* et *Z.*, Jap. *Tama-kusu, Yama-kusu, Madami;* the bark of this plant (553. b.) serves as a brown dye in *Hachijō-jima*.

374. c. Ternstroemia japonica, *Th,,* Jap. *Mokkoku;* the bark of this plant (661) is used as a brown dye for clothes in the Islands of *Hachijō, Miyake* and *Mikura*.

374. d. Rhaphiolepis japonica, *S.* et *Z.*, Jap. *Hama-mokkoku, Hebaru-no-ki* (*Satsuma*), *Tekachigi* (*Loochoo*), *Saema* (*Hachijō-jima*); the bark of this plant (671) is used for dying dark brown. In the islands of *Okinawa, Oshima* and *Hachijō*, it is used to dye *Tsumugi* and other clothes.

374. e. Bruguiera gymnorrhiza, *Lamk.*, Jap. *Taka-tsuku, Kiire-tsuku, Hiroki* (*Ōshima*); an evergreen shrub of the order Rhizophoraceæ growing in the sea-coasts of *Satsuma*. The bark of this plant is collected and used to dye reddish brown. Its use is the same as the imported Rhizophora bark.

375. Quercus dentata, *Th.*, Jap. *Kashiwa;* a deciduous tree of the order Amentaceæ growing wild in cold regions, attaining to a height of 20-30 fts. After the new leaves come forth, it produces male and female flowers separately, the male in an ament form and the female like that of an acorn. The kernel of the acorn is bleached and eaten as food. The dried bark is rich in tannin and used mostly to dye fishing nets with the name of *Kashiwagi*. There are other species of the Quercus, as Quercus cuspidata, Q. serrata, and Q. glandulifera, which are used for the same purpose and also for tanning leather. There are two kinds of *Kashiwa*. One with thin narrow leaves, which fall in winter, is called *Nara-kashiwa*, and the other with thick broad leaves and remaining long on the branches after withered is called *Mochi-kashiwa*. The latter leaves are used to wrap cakes.

375. b. Elaeacocca cordata, *R. Br.*, Jap. *Abura-giri;* the bark of this plant (313) is used for dying as in the preceding.

376. Alnus maritima, *Nutt.*, var. japonica, *Regel.*, Jap. *Han-no-ki;* a deciduous tree of the order Amentaceæ growing wild in wet places, attaining to a hight of 20-30 fts. In spring it produces male and female flowers separately before it sprouts. The male flowers hang down from the branches in the form of a catkin, and the female yield round fruits with scales. In autumn when the fruits fully ripen, being about 1 inch in length, they are collected and dried for dying. Other trees of this genus, as the *Mehari-no-ki, Yama-han-no-ki, Nikkō-bushi*, and *Kawara-han-no-ki*, have the same use.

376. b. Alnus maritima, *Nutt.*, var. obtusa, *Fr. et Sav.* Jap. *Mehari-no-ki;* a variety of the preceding with round leaves. Its fruits have the same quality as the preceding.

377. Alnus incana, *Willd.*, var. glauca, *Ait.*, Jap. *Yama-hannoki ;* a deciduous tree of the order Amentaceæ growing wild in mountainous regions. It resembles 376, attaining to a height of about 10 fts., with broad dissected leaves. The fruits are large and of a better quality. Fishing nets are coloured with the juice of this bark, whence it is called *Ami-kawa* (net-bark).

378. Alnus firma, *S.* et *Z.*, var. multinervia, *Reg.*, Jap. *Nikko-bushi, Yasha-bushi ;* a deciduous wild tree of the order Amentaceæ. It is mostly small, but some one grows about 10 fts. high. the fruits are oval resembling those of 376, but are of a larger size and more useful, being used instead of gall.

378. b. Alnus viridis, *Dc.*, var. sibirica, *Regel.*, Jap. *Kawara-hannoki ;* a deciduous tree of the order Amentaceæ growing plentifully on the *Fuji* mountain. It resembles much the preceding in shape, but it has broader leaves and smaller fruits. The fruits are esteemed in dying silk in *Kai* province.

378. c. Platycarya strobilacea, *S.* et *Z.*, Jap. *Nobu-no-ki, Nogurumi ;* a deciduous tree ef the order Juglandaceæ growing in warm countries, attaining to a height of 10 fts. Barren and fertile flowers appear at the same time with the leaves, and thorny fruits are produced. The bark is used for dying fishing nets.

379. Curcuma longa, *L.*, var. macrophylla, *Miq.*, Jap. *Ukon ;* a perennial herb of the order Zingiberaceæ found in warm regions, sprouting in spring. The leaves are about 2 fts. long, and in summer flowers appear in cluster. In autumn the roots are collected for a yellow dye and medicine.

379. b. Curcuma longa, *L.*, Jap. *Kyō-ō ;* a species of the preceding with the same form and use.

280. Commelina benghalensis, *L.*, Jap. *Ōboshibana ;* a biennial plant of the order Commelinaceæ cultivated in *Yamada* in the province of *Ōmi*. It is sown in autumn and transplanted in spring. It attains to a height of 3-4 fts. It bears blue flowers in summer and autumn. They are picked every morning and pressed to papers, which are called *Aigami* (indigo paper). The colour is very fine, but liable to fade.

381. **Miscanthus chrysantes**, *Max.*, Jap. *Kari-yasu ;* a perennial wild grass growing about 3 fts. high. The panicles resemble those of Eularia, but are mostly divided into 3 parts. They are cut, dried, and preserved as a yellow dye.

382. **Arthraxon ciliare**, *Beauv.*, Jap. *Kobuna-gusa*, *Hachijō-kariyasu ;* a perennial wild grass. Its fine procumbent stems creep over the ground, stand upright at the end, and bring forth panicles in several divisions. Formerly this herb was used as a yellow dye, but now only used in the *Hachijō* iceland to dye silk.

383. **Rhus semialata**, *Murray.*, var. osbeckii., *Dc.*, Jap. *Nurude, Fushi-no-ki, Katsu-no-ki ;* a deciduous wild tree of the order Anacardiaceæ growing to a height of about 10 fts. In summer, it yields fine little flowers disposed in panicles, being succeeded with small fruits. Small insects come and stain the fruits with salt-like white powder. Thus the upper or lower surface of the leaves swells up, and finally brown galls rich in tannin are formed. The galls are much esteemed for dying and other numerous purposes.

CHAPTER XIII—Odorous Plants.

This Chapter includes those plants, the flowers of which are esteemed for their fragrance, being used to put them in scent bags or to give water their odour. There are also included those which have fragrant fruits, leaves, stems, or roots, but those described in the chapter of condiments and spices are excluded here.

384. **Magnolia kobus**, *Dc.*, Jap. *Kobushi ;* a deciduous wild tree of the order Magnoliaceæ, growing 1-2 fts. high. In'

spring it bears flowers before leaves. The flowers are single petaled and white shaded with pink, having a nice odour.

385. Rosa banksiæ, *R. Br.*, Jap. *Mokkō-bara;* a deciduous climber of the order Rosaceæ, thriving well in gardens. In spring it bears buds with leaves, and yields double white fragrant flowers in summer.

386. Rosa, Jap. *Hakuō-bara;* a deciduous shrub of the order Rosaceæ. When kept in hot houses the leaves do not fall as an evergreen. It bears fragrant yellowish white flowers all the year.

387. Rosa multiflora, *Th.*, Jap. *No-ibara;* a deciduous wild shrub of the order Rosaceæ. Its long slender branches grow like vines. In summer it bears 5 petaled white fragrant flowers in the form of a raceme. Its red berries resemble those of Nandina, and are used for medicine, being called *Yé-jitsu*. This tree varies much in size. The flowers are pink or bright red, and very beautiful.

388. Rosa luciæ, *Fr.* et *Sav.*, Jap. *Teriha-no-ibara;* a species very much like the preceding in shape and quality, with smaller lustrous leaves, later opening and more odorous flowers, and larger fruits.

389. Chimonanthus fragrans, *Lindl.*, Jap. *Rōbai, Nankin-mume;* a deciduous shrub of the order Calycanthaceæ planted in gardens, growing about 10 fts. high. In winter it bears buds and begins to bloom in December. The flowers give a nice odour till the end of February. The inner petals are of a mottled purple colour, and the outer ones are large and have a yellow waxy appearance, whence the name *Rōbai* (wax-plum). There are several varieties. The one here mentioned is *Shinno-Robai*.

390. Jasminum grandiflorum, *L.*, Jap. *Sokei;* an evergreen shrub of the order Jasminaceæ introduced from foreign countries about the year 1820. As it is the product of warm regions and cannot bear cold climates, it must be kept in hot houses. The stem is soft and flexible like a vine, attaining to a height of about 4 fts. In early summer it brings forth panicles of single white fragrant flowers.

391. Jasminum sambac, *Ait.*, Jap. *Mōrinkwa ;* an evergreen shrub of the order Jasminaceæ introduced from Loochoo Islands about the year 1596. The stems attain to a height of about 4 fts. Late in summer, it bears single white flowers with a very strong agreeable odour. It resembles the former being of the same genus, and in winter it must be kept in hot houses.

392. Olea fragrans lutea, Jap. *Mokusei, Kin-mokusei ;* an evergreen tree of the order Oleaceæ planted in gardens, attaining to a height of about 10 fts. Late in autumn, it produces small reddish yellow flowers in clusters from the axils of the leaves. The flowers are strongly fragrant and called *Tankei* (red olea fragrans).

393. Olea fragrans alba, Jap. *Gin-mokusei ;* a variety of the former of the same quality with white flowers and larger leaves, but with less odour. A variety having leaves with coarsely dentated edges is called *Hiragi-mokusei*.

394. Daphne odora, *Th.*, Jap. *Jinchō-ge ;* an evergreen shrub of the order Thymelaeaceæ planted in gardens, attaining to a height of 3-5 fts. It grows in the form of a thicket. In winter it brings forth buds in clusters, which open in spring. The flowers are of purplish red outside, and white inside. The leaves are marginned with pale yellow. That with white flowers has leaves not marginned, and grows higher than the other, with a stronger odour. The one here illustrated refers to the latter kind.

395. Cymbidium ensifolium, *Sw.*, Jap. *Suruga-ran, O-ran ;* an evergreen herb of the order Orchidaceæ growing wild in the provinces of *Iyo, Kii,* and especially *Suruga,* whence the name. The leaves are 2-3 fts. long. In summer and autumn it bears pale yellow flowers shaded with green. As they have a fragrant odour, it is planted in pots.

396. Cymbidium, Jap. *Hōsai-ran ;* an evergreen terrestial orchid growing wild in mountains of warm countries. The leaves are of a lustrous dark green, and 2-3 fts. long and about an inch broad. In spring dark purplich red fragrant flowers are produced on a stalk.

397. Cymbidium, Jap. *Kan-ran ;* an evergreen terrestial orchid growing wild in mountains of warm regions, differing in size according to the its growing places. It blooms in the beginning of winter, whence the name *Kan-ran* (winter orchid). One with yellowish green flowers illustrated here is called *Sei-kan-ran* (green winter orchid). Another with a purplish red shade is called *Shi-kan-ran* (purple winter orchid). There are manyo ther varieties, all of which have fragrant flowers and are admired for blooming in winter.

Note.—Those mentioned here are only a very few of the flower bearing plants. There are a great many others used for making perfumed water and oil.

CHAPTER XIX.—MEDICINAL PLANTS.

This chapter includes those plants, which are grown wild or cultivated in this country and used for medicine. There are a great many medicinal plants, but those mentioned here are the most noted. The poisonous plants are included in their own chapter, and they are not mentioned here.

398. Coptis brachypetala, *S.* et *Z.*, var. major, *Miq.*, Jap. *Ōren, Seriba-ōren ;* a perennial plant of the order Ranunculaceae growing wild in mountains. In spring it shoots forth stalks by the sides of the leaves to a height of 3-5 inches, and yields small white flowers on divided petioles. When young leaves flourish, the old ones die. The roots are taken and dried for a medicine and a yellow dye. They have a very bitter taste.

399. Coptis occidentalis, *Nutt.*, Jap. *Kikuba-ōren ;* a species of the former, with leaves resembling those of Chrysanthemum, having the same quality and being used as the former.

400. Coptis trifolia, *Salisb.*, Jap. *Mitsuba-ōren ;* a species of Coptis with ternate leaves of the same quality and use as 398.

401. Coptis brachypetala, *S.* et *Z.*, Jap. *Hosoba-ōren ;* a species of Coptis with small fine leaves having the same quality and use as 398.

402. Schizandra chinensis, *Baill.*, Jap. *Chōsen-gomishi ;* a deciduous climbing shrub of the order Magnoliaceæ brought from Corea about the year 1717. In spring it brings forth thin petioles bearing flowers in small panicles, and yields red fruits, which are dried and used as a medicine. The vine has an agreeable odour,

402. b. Schizandra nigra, *Max.*, Jap. *Matsubusa, Ushibudō ;* a variety of the former growing wild in mountains, having the same quality and use.

403. Kadsura japonica, *L.*, Jap. *Binan-kadsura, Binan-sō, Sane-kadsura ;* an evergreen climbing plant of the order Magnoliaceæ, grown wild in mountains and also planted in gardens. In summer it shoots forth a thin and short petiole with many flowers, and each petiole has many small red berries accumulated on a globular stock. The berries are dried for medicine. As the vine is rich in a mucilaginous fluid, it is dried and used for paper making or hair dressing.

404. Corydalis ambigua, *Cham. Schlecht.*, Jap. *Tsubute ;* a perennial herb of the order Fumariaceæ brought from China about the year 1720. There are 2 varieties with small and large leaves, and also a variety of Japanese origin. In spring it grows about 5 inches high, and it blooms in March and April. When the leaves wither in May, the roots form small tubers, which are dried for medicine.

405. Polygala sibirica, *L.*, Jap. *Hime-hagi;* an evergreen wild herb of the order Polygalaceæ. Several stalks, coming forth from one root, attain to a height of $\frac{1}{2}$-1 ft. and generally lie on the ground. In summer, they bloom purple flowers of papilio-shape in the leaf-axils. The fruits are flat and round, being about $\frac{1}{3}$ inch in diameter. The roots are used as a medicine.

406. Malva pulchella, *Bertin.*, Jap. *Fuyu-aoi, Kan-aoi;* a biennial herb of the order Malvaceæ, growing wild on the sea-coasts of many provinces, and also planted in gardens. The stem attains to a height of 3-5 fts., and from spring to winter it bears flowers in clusters on the axils of leaves. The flowers are about $\frac{1}{2}$ inch in diameter, and are yellowish white with purple shade. The fruits are used as a medicine after drying. The young leaves are eaten. One variety with shrivelled edges is called *Okanori*.

407. Orixa japonica, *Th.*, Jap. *Kokusagi;* a deciduous wild shrub of the order Rutaceæ, growing about 10 fts. high. It is a diœcious plant. The leaves are smooth and lustrous, and have a strong disagreeable odour. Late in spring, it yields flowers on the axils of the leaves. The male flowers form a panicle. The female flowers are 4 pataled and yellow coloured. producing small fruits. The roots are used as a medicine by drying.

408. Rhamnus japonica, *Max.*, Jap. *Kuro-mume-modoki;* a deciduous wild shrub of the order Rhamnaceæ, growing 6-7 fts. high. It is provided with thorns. In summer it bears greenish yellow flowers on the axils of the leaves. The fruits are round and black, serving as a purgative after drying.

409. Zizyphus vulgaris, *Lam.*, var. inermis, *Bunge.*, Jap. *Sanebuto-natsume;* a deciduous tree of the order Rhamnaceæ, brought from China about the year 1717, attaining to a height of about 10 fts. The branches are provided with sharp thorns. The shape resembles that of 188. The fruits are round and aciduous, and edible when turn red by ripening. The seeds are crushed and their pernels are used for medicine.

410. **Astragalus reflexistipulus,** *Miq.,* Jap. *Momendsuru ;* a perennial herb of the order Leguminosæ growing in mountains. The stem lies on the ground in the form of a vine. In summer yellow or purple flowers are produced in the axils of leaves, being succeeded with pods. The roots are dried and used for medicine.

411. **Trigonella foenum-graecum,** *L.,* Jap. *Koroha ;* an annual cultivated plant of the order Leguminosæ, brought from China about the year 1717. It is sown in spring, growing 2-3 fts. high. In summer it produces small white papilionaceous flowers from the axils of the leaves, being succeeded with thin pods of about 3 inches in length. The ripen seeds are used for medicine.

412. **Glycyrrhiza echinata,** *L.,* Jap. *Amakusa ;* a perennial leguminous herb brought from China about the year 1717. In spring it sprouts, growing 2-3 fts. high. In late summer light purple papilionaceous flowers are produced from the axils of leaves. When the stems are dead, the roots are dug out and dried. The dried roots are yellow and sweet, and are used for medicine or for mixing to food.

413. **Euchresta japonica,** *Benth.,* Jap. *Miyama-tobera, Isha-daoshi ;* an evergreen leguminous climbing shrub growing in shady places in mountains of warm regions. The stem is tender, 1-2 fts. long, and liable to lie on the ground. In summer white popilionoceous flowers are produced in panicles from the axils of leaves. The seeds are black when ripen. The roots are dried for medicine.

414. **Sophora angustifolia,** *S.* et *Z.,* Jap. *Kurara ;* a perennial wild herb of the order Leguminosæ. It produces several stems from one root, and grows 3-4 fts. high. Light yellow popilionaceous flowers are produced in panicles at the head

of branches, being succeeded with long pods. The roots are used for medicine. The juice obtained by boiling the stems and leaves is used to destroy insects injurious to vegetables. Fibre is taken from the bark of the stems.

415. Psoralea corylifolia, *L.*, Jap. *Oranda-hiyu ;* an annual leguminous plant brought from China about the year 1717. It is sown in spring, and grows 3-4 fts. high. In late summer, small balls of light purple flowers are produced on short stalks from the axils of leaves. The ripen black seeds are used for medicine.

416. Geum japonicum, *Th.*, Jap. *Daikon-sō ;* a perennial wild herb of the order Rosaceæ, growing 2-3 fts. high. In summer it bears 5 petaled deep yellow flowers at the head of the stems, being succeeded with prickly balls, which are about half inch in diameter and contain much seeds. The roots are dried for medicine. The young plants are eaten as a vegetable.

417. Poterium officinale, *L.*, Jap. *Waremokō ;* a perennial wild herb of the order Rosaceæ attaining to a height of 3-4 fts. In autumn, it bears groups of small purplish red flowers at the top of the young branches. The reare other varieties, which flowers are pink, crimson or white. The roots of the common variety are used as medicine by drying.

418. Bupleurum falcatum, *L.*, Jap. *Kamakura-saiko;* a perennial wild herb of the order Umbelliferæ, growing 2-3 fts. high. In autumn it bears small yellowish flowers in clusters at the head of the branches from the axils| of leaves. The roots are gathered in winter and used as medicine.

419. Foeniculum vulgare, *Gærtn.*, Jap. *Kureno-o-mo, Uikyō ;* a biennial herb of the order Umbelliferæ cultivated in fields. In summer the stem attains to a height of 6-7 fts, dividing into many branches, which bring forth small yellowish flowers in clusters at the head. When the seeds ripen the plants die. The seeds have a strong agreeable odour, and they are used as a medicine. An oil is also extracted from them.

420. Anethum graveolens, *L.*, Jap. *Inondo, Himenikyō;* a biennial plant of the order Umbelliferæ, resembling the former in shape, but smaller, being about 2 fts. high. Its odour is not so strong, but it has almost the same use as the former.

421. Selinum japonicum, *Miq.*, Jap. *Hamazeri, Hamaninjin;* a biennial herb of the order Umbelliferæ, growing wild in the sandy places of sea coasts, attaining to a height of 5-6 inches. In autumn small white flowers come forth in clusters. After the seeds are ripen, the plants die. The seeds are collected and used as a medicine.

422. Ligusticum acutilobum, *S. et Z.*, Jap. *Tōki, Yamazeri;* a perennial herb of the order Umbelliferæ, growing wild in mountainous regions, and also being cultivated in gardens. In summer it attains to a height of 2-3 fts., and brings forth small white flowers in clusters at the head of the branches. The roots are collected and used as a medicine.

423. Silar divaricatum, *Benth. et Hook.*, Jap. *Fudebōfū;* a triennial herb of the order Umbelliferæ brought from China in the year 1717, attaining to a height of 2-3 fts. In summer it bears small white flowers in clusters at the head of the stem. After the seeds are ripen, the roots die. The roots are collected in the autumn of the second year and dried for medicine.

424. Archangelica gmelini, *Dc.*, *Shishi-udo;* a biennial wild herb of the order Umbelliferæ. The stems and leaves are covered with coarse hair. In summer the stem grows to a height of 6-7 fts., and in autumn it brings forth small white flowers in clusters. After the seeds are ripen, the roots die. The roots are taken in winter and dried for medicine.

425. Angelica anomala, *Pall.*, Jap. *Yoroi-gusa;* a triennial herb of the order Umbelliferæ cultivated in gardens.

The chinese kind was introduced to this country in the year 1717. The stem attains to a height of 7-8 fts. In summer it bears small white flowers in an umbel at the top of the stem, and after the seeds are ripen the plant dies. The Roots are collected in winter and dried for medicine.

426. Angelica decursiva, *Miq.*, Jap. *Nodake, Manzairaku;* a triennial wild herb of the order Umbelliferæ, growing 7-8 fts. high. In autumn it bears small dark purple or white flowers in an umbel. The roots are collected and used for medicine by drying.

427. Coriandrum sativum, *L.*, Jap. *Koyendoro;* a biennial herb of the order Umbelliferæ, introduced from a foreign country. It is sown in autumn, and grows to a height of 1-2 fts. in the following year. In late summer it bears small flowers in an umbel, and yields fragrant seeds to be used as medicine. The leaves and stems have a slight disagreeable odour. The large foot leaves are eaten as a vegetable.

428. Conioselinum univittatum, *Turcz.*, Jap. *Senkiō, Omuna-kadsura;* a biennial herb of the order Umbelliferæ growing wild and also cultivated in gardens. The stems attain to a height of 1-2 fts., and in autumn they bring forth small yellowish white flowers at the top. The roots have strong fragrant odour and are used as medicine.

429. Angelica ?, Jap. *Udo-modoki;* a biennial wild herb of the order Umbelliferæ. Resembling 424 in shape, the under side of the leaves is nearly white, the stems and leaves have no hair, and the stems are purple. In summer the stems grow to a height of 3-4 fts. and bear small white flowers. The roots are haraested for medicine.

430. Panax repens, *Maxim.*, Jap. *Tochiba-ninjin;* a perennial herb of the order Araliaceæ growing wild in shady places of mountains. After 3 pears the stems grow to a height of about 2 fts., being divided into 3 branches with 5 cleft leaves, and bearing small 5 petaled white flowers in cluster at the top of the branches. In autumn small fruits are ripen and beautifully red. The roots have knots, though there are straight roots. The roots are dried for medicine.

431. Panax ginseng, *C. A. Mey.*, Jap. *Ninjin, Kano-nige-kusa ;* a perennial herb of the order Araliaceæ introduced and cultivated. In third year after sowing, the stems grow about 2 fts. high, and bear flowers and seeds. It resembles the preceding in shape, and the main roots are large. The roots are steamed and dried for medicine, and much exposted to China.

432. Cornus officinalis, *S.* et *Z.*, Jap. *Sanshuyu ;* a deciduous tree of the order Cornaceæ, growing about 10 fts. in fields. It bears small fine yellow flowers in cluster before sprouting in spring. In autumn the oblong red aciduous fruits are collected and dried for medicine.

433. Sambucus racemosa, *L.*, Jap. *Niwatoko, Kitadsu;* a deciduous shrub of the order Caprifoliaceæ growing wild or planted in gardens, attaining to a height of 10 fts. Late in spring it bears small white flowers in an umbel, forming small round red or yellow fruits. The stems and flowers are dried for medicine. A good edible fungus called *Kikurage* grows on the rotten stem.

434. Lonicera confusa, *Dl.*, Jap. *Suikadsura ;* a deciduous wild climber of the order Caprifoliaceæ. In the beginning of summer it produces purplish white fragrant flowers on the axils of leaves. The flowers gradually turn yellow as they become old, and they are dried for medicine. It produces black round fruits. The vines and leaves are dried and used instead of tea.

435. Uncaria rhychophylla, *Miq.*, Jap. *Kagikatsura ;* an evergreen climbing plant of the order Rubiaceæ, growing wild in warm regions. Late in autumn it bears small light brown flowers in the form of a ball about an inch large. The leaves grow opposite, and a hooked spine grows in the root of each leaf. The spines are used for medicine.

436. Valeriana officinalis, *L.*, Jap. *Kanoko-sō, Haruominayeshi;* a perennial herb of the order Valerianaceæ growing wild in mountains. In spring its stem grows to a height of 1½ fts., bearing small pink flowers in an umbel. The roots are dried and used as medicine. Their smell is too strong.

437. Inula japonica, *Th.*, Jap. *Oguruma;* a perennial wild herb of the order Compositæ, growing to a height of 2-3 fts. The divided branches bear yellow single petaled flowers, which are used for medicine by drying. There are also those with double or tubular flowers, being planted in gardens for their beauty.

438. Artemisia capillaris, *Th.*, Jap. *Kawara-yomugi;* a perennial herb of the order Compositæ, growing wild in sandy places near rivers. The leaves are soft, slender and beautiful. In summer, its stem attains to a height of 2-3 fts. The branches bear many fine flowers which are stronger than *Yomogi* (68) in flavour. The seeds are used for medicine.

439. Atractylis lyrata, *S. et Z.*, Jap. *Okera;* a perennial herb of the order Compositæ, growing wild, and also planted in fields. The stem attains to a height of 2-3 fts. In autumn it bears white flowers in the form of a ball. A variety called *Biyaku-jutsu* have long and narrow leaves and purple flowers. The roots of the 2 varieties are large and have many rootlets. They are dried and used for medicine.

440. Rehmannia lutea, *Max.*, Jap. *Sao-hime;* a perennial herb of the order Cyrtandraceæ. In spring it grows to a height of 7-8 inches. In early summer it bears yellowish white flowers shaded with purple on the divided branches at the top of the stems. Before the fruits ripen the stems die. The roots are used for medicine by drying directly or after steaming.

441. Ophelia diluta, *Ledeb.*, Jap. *Senburi;* an annual wild herb of the order Gentianaceæ, growing about 1 ft. high. In autumn several flowers appear at the top of the stems. The flowers r ae5 petaled and pale red shaded with purple. There is a variety with large leaves, which is pictured. The stems and leaves are dried and used for medicine.

442. Endotropis caudata, *Miq.*, Jap. *Ikema;* a perennial climber growing in mountains. In spring the vines are produced, bearing small white flowers which are succeeded with capsules. In autumn the capsules split out white fibre. The tubular roots are collected and dried for medicine.

443. Scrophularia oldhami, *Oliv.*, Jap. *Goma-kusa;* a perennial wild herb of the order Scrophulariaceæ. In summer the stems grow 1-5 fts. high, and produces light yellow flowers in a small panicle. The large roots are used for medicine.

444. Nepeta japonica, *Max.*, Jap. *Keigai;* an annual herb of the order Labiateæ. It is sown in spring, growing about 2 fts. high in summer. Reddish white small flowers are produced in a panicle at the top of the stem. When the seeds ripen the plant dies. The seeds are fragrant and used for medicine.

445. Scutellaria machrantha, *Fisch.*, Jap. *Koganeyanagi, Kogane-bana ;* a perennial herb of the order Labiateæ. In summer the stem grows about 2 fts. high, and bears white or purple flowers in panicles. The large deep yellow roots are dried for medicine.

446. Mentha arvensis, *L.*, var. vulgaris, *Benth.*, Jap. *Hakka, Mekusa ;* a perennial herb of the order Labiateæ mostly cultivated. It sprouts in spring, and grows about 1 ft. high in summer, opening small purple labiate flowers. The stems and leaves are dried and used for medicine or for taking the oil, which is very fragrant and refreshing.

447. Vitex trifolia, *L.*, var. unifoliolata, *Schauer.*, Jap. *Hamagō, Hamashikimi ;* a deciduous shrub of the order Verbenaceæ, growing on sea-coasts 3-4 fts. high. The branches creep over the ground like vines. In summer it bears dark purple labiate flowers disposed in panicles at the top of the stem. The small round fragrant seeds are dried and used for medicine.

448. Plantago asiatica, *L.,* Jap. *Ōbako;* a perennial herb of the order Plantaginaceæ, growing wild everywhere. It shoots forth flower stalks of several inches in length from the centre of the leaves. The seeds are dried and used for medicine. The young leaves are eaten as a vegetable.

449. Celosia argentea, *L.,* Jap. *No-geitō, Fude-keito;* an annual herb of the order Amarantaceæ, growing wild and also planted in gardens. It attains to a height of 1-2 fts. The flowers are light red and have the form of a Japanese pen. The seeds are used for medicine.

450. Achyranthes bidentata, *Bl.,* var. japonica, *Miq.,* Jap. *Inokodsuchi;* a perennial herb of the order Amaranthaceæ, growing wild everywhere, attaining to a height of 2 fts. with square stems. In summer it bears flowers in panicles, and forms small thorny seeds which easily attach to clothes. The roots are used for medicine by drying.

451. Polygonum aviculare, *L.,* Jap. *Niwa-yanagi;* an annual wild herb of the order Polygonaceæ. In spring its stem attains to a height of 10 fts., and in summer it bears small flowers. Its stems and leaves are used for medicine.

452. Polygonum multiflorum, *Th.,* Jap. *Tsurudokudami;* a perennial wild climbing plant of the order Polygonaceæ. In autumn it bears small white flowers in panicles. The roots consist of many large tubers, and are used for medicine by drying.

453. Rheum undulatum, *L.,* Rhubarb, Jap. *Ohoshi;* a perennial herb of the order Polygonaceæ. It sprouts in spring, and in summer the stem grows to a height of 3-6 fts., bearing flowers. Its roots are dried for medicine, and its petioles are used as a vegetable.

454. Cinnamomum loureirii, *Nees.,* Jap. *Nikkei;* an evergreen tree of the order Lauraceæ, growing 20-30 fts. high In summer it bears small yellowish green flowers. The bark of the branches and roots are dried and used for medicine, being aromatic and refreshing.

455. Daphnidium strychnifolium, *S.* et *Z.*, Jap. *Uyaku ;* an evergreen shrub of the order Lauraceæ, growing 4–9 fts. high in the form of a bush. It bears yellowish green flowers in bunches in the axils of leaves, and produces small red berries. The tubular roots are dried and used for medicine.

456. Asarum sieboldi, *Miq.*, Jap. *Hiki-no-hitai-gusa ;* a perennial herb of the order Aristolochiaceæ, growing wild in shady places in mountains. In spring it sprouts and bears a dark purple flower of ½ inch in size near the ground. The rootlets are dried and used for medicine.

456. b. Asarum variegatum, *Al.*, Jap. *Kan-aoi ;* a species of the former growing wild in shady places in mountains The leaves are round and pointed at the tops, and concave near the petioles resembling a horse hoof. The leaves are long or round, and differ also in size. The rootlets are used instead of the former, but they are inferior, being acrid and bad smelled.

457. Aristolochia kæmpferi, *Willd.*, Jap. *Uma-no-sudsu, Ohaguro-bana ;* a perennial climbing herb of the order Aristolochiaceæ growing wild. In summer it shoots forth long stalks at the axils of the leaves, and yields tubular flowers on small balls. The flowers open at the tip, and form purplish green petals. The roots are dried and used for medicine. There are other varieties with large or narrow leaves and with white flowers.

458. Houttuynia cordata, *Th.*, Jap. *Doku-dami, Jō-yaku ;* a perennial herb of the order Saururaceæ, growing wild and attaining to a height of 7–8 fts. It gives a disagreeable odour when touched. In summer it bears 4 petaled white flowers of different sizes. The roots are dried and used for medicine.

459. Dioscorea sativa, *L.*, var. rotundis, *Fr.* et *Sav.*, Jap. *Tokoro ;* a perennial wild climbing herb of the order Dioscoreaceæ. It resembles very much D. japonica (111) in form, though its vine turns left. The fibrous roots are used for medicine by drying. The tubers are eaten steamed, and are also used to make starch. Another kind with lobed leaves is bitter and can not be eaten.

460. Heterosmilax japonica, *Kuuth.*, Jap. *Sankirai;* an evergreen climbing herb of the order Smilaceæ. It is a diaecious plant. In spring it shoots forth vines, with hooked tendrils under each leaf and with about 10 purplish green flowers on the axils of leaves, being succeeded with small round black fruits. The tubular roots are used for medicine when dried. There are several varieties with narrow or round leaves.

461. Stemone japonica, *Miq.*, Jap. *Hodotsura;* a perennial herb of the order Roxburghiaceæ. There are two kinds, standing and climbing. The variety here mentioned is the climbing one. In summer it yields one or two flowers in the centre of the leaves. Many oblong small tubers attached to the roots are dried and used for medicine.

462. Gastrodia elata, *Bl.*, Jap. *Nusubito-no-ashi*, *Kami-no-yagara;* a parasite of the order Orchidaceæ growing wild in mountains. In early summer it shoots forth a straight yellowish red stem to a height of 4-5 fts., bearing flowers in panicles at the head. In autumn the stems and roots die. The tubers growing laterally to a length of about 10 inches with a diameter of an inch are used for medicine.

463. Curcuma longa, *L.*, Jap. *Kyō-ō*, *Haru-ukon;* a perennial herb of the order Zingiberaceæ. It resembles *Ukon* (379), with white hair under the leaves. Late in spring, it shoots forth stalks to a height of 6-8 inches, bearing 2 yellow flowers. The tuberous roots are dried and used for medicine. They have the smell of ginger, and are yellow.

464. Amomum, Jap. *Gajutsu*, *Usuguro;* a perennial herb of the order Zingiberaceæ. It resembles the preceding in shape, with dark purple variegation in the centre of the leaves.

In summer it bears red flowers. The dark green tubers are dried and used for medicine.

465. Alpinia japonica, *Miq.*, Jap. *Hana-myōga;* an evergreen herb of the order Zingiberaceæ growing wild in warm regions. It resembles *Myōga* (128) in form. In summer it yields white flowers shaded with pink in panicles, being succeeded with bean-sized red berries containing numerous small seeds, which are used for medicine.

466. Anemarrhena asphodeloides, *Bunge.*, Jap. *Hana-suge;* a perennial herb of the order Liliaceæ cultivated in fields. It grows in bushes with narrow long leaves. In summer it shoots forth a flower stalk to a height of about 2 fts., and bears flowers in panicles. Its rhizomes are used for medicine.

467. Ophiopogon spicatus, *Gawl.*, Jap. *Yabu-ran;* an evergreen herb of the order Liliaceæ growing wild in woods. In summer its stalk grows to a height of about 1 ft. and bears purplish red flowers in panicles, being succeeded with black bean-sized round fruits. The small globular tubers are used for medicine.

468. Asparagus lucidus, *Lindl.*, Jap. *Kusasugi-kadsura;* a perennial herb of the order Liliaceæ growing wild on sea-coasts and also cultivated in fields. There are standing and climbing varieties. In summer it produces small yellowish flowers, which are succeeded with little red berries. The tuberous roots grow in tufts and are used for medicine or preserved in sugar.

469. Lemna polyrhiza, *L.*, Jap. *Uki-kusa;* an annual aquatic herb of the order Naiadaceæ growing on the stagnant water surface. A variety has entirely green leaves and 2 fibrous roots. Another variety has leaves which are green on the upper and purple on the under side, and is provided with several large fibrous roots. Both are dried and used for medicine.

470. Cyperus rotundus, *L.*, Jap. *Hama-suge;* a perennial herb of the order Cyperaceæ growing wild especially near sea-shores. In summer it bears flowers at the head of the stalk, which grows to a height of about 1 ft. In winter its small tuberous roots are collected, dired, and used for medirine.

471. Scirpus maritimus, *L.*, Jap. *Mikuri;* a perennial aquatic grass of the order Cyperaceæ resembling the preceding in form, growing to a height of 4–5 fts. Its tuberous roots covered with black hair are dried and used for medicine.

472. Pachyma cocos, *Smi.*, Jap. *Matsuhodo;* a parasitic fungus growing on pine-roots under ground, forming a tuber about the size of a baby's head. The outer skin is black and wrinkled, but the interior is white or light pink.

473. Boletus laricus, *Linn.*, Jap. *Eburiko;* a parasitic fungus growing on the old stems of Larix leptolepis in the mountains of northern provinces. It forms a white brittle tuber of 6–7 inches, being used for medicine.

474. Jap. *Meshimakobu;* a parasitic fungus growing on the old stems of mulberry trees found in the island of *Meshima*, one of 5 Islands in *Hizen*. Its outside is brown, while the inside is yellow. It is used for medicine.

475. Lycoperdon boviste, *L.*, Jap. *Hokoritake, Chidome, Mimitsubushi;* a terrestial fungus growing in shady places in mountains in autumn. It is a small ball about the size of a man's head, being dark brown and cotton-like. When touched there arises from it smoke-like powder, which is used to stop bleeding.

476. Jap. *Hagi-hodo;* a fungi growing under ground, especially found in the province *Tanba*. It resembles 472 in shape, but its tuber is smaller. The outer-skin is black or red, and the inner grayish white. It is dried and used for medicine.

477. Digenea wulfeni, *Kg.*, Jap. *Makuri;* an alga growing on rocks in the sea of southern provinces. It is slender and divided into branches which are about ½ ft. high. It is green and rough, and it is used for medicine by drying.

CHAPTER XX.—POISONOUS PLANTS.

This Chapter includes the plants which are poisonous, though some are used for medicine. As they are poisonous, care must be taken of those which are growing wild.

478. Clematis paniculata, *Th.*, Jap. *Senninsō, Hakobore, Takatade ;* a perennial climbing herb of the order Ranunculaceæ growing wild. In autumn it bears 4 petaled white flowers, being succeeded with fruits of hairy balls. The leaves and stems contain a poisonous ingredient. When chewed, it hurts teeth, whence the name of *Hakobore* (teeth-breaker). It also blisters the skin when touched.

479. Ranunculus acris, *L.*, Jap. *Kinpōge ;* a perennial herb of the order Ranunculaceæ growing wild. In spring the stem grows to a height of 1-2 fts., with 5 petaled yellow or white flowers, which are sometimes doubled. It contains a narcotic ingredient.

480. Ranunculus sceleratus, *L.*, Jap. *Tagarashi, Tatarabi ;* a biennial aquatic herb of the order Ranunculaceæ In spring the stem grows to a height of 1-2 fts., with very lustrous leaves and 5 petaled small yellow flowers. It has the same form and quality as the preceding.

481. Ranunculus ternatus, *Th.*, Jap. *Kitsune-nobotan ;* a biennial herb of the order Ranunculaceæ growing wild in moist ground. The leaves are parted and covered with hair. The stem attains to a height of about 1½ fts. and bears small yellowish flowers. It has the same quality and form as the preceding.

482. Aconitum chinense, *S.* et *Z.*, Jap. *Kabuto-giku, Kabutosō, Torikabuto ;* a perennial herb of the order Ranunculaceæ planted in gardens for its flowers, the tuberous roots being used for medicine. In spring it grows to a height of about 2 fts. In autumn it blooms many blue purple or white helmet-shaped flowers.

483. Aconitum fischeri, *Reichenb.*, Jap. *Yama-tori-kabuto ;* a species of the preceding growing wild in mountains. In form it is much alike, growing to a height of 3–4 fts. The colour of the flowers is deep purple or blue. It has also the same quality as the former.

484. Aconitum uncinatum, *L.*, var. japonicum, *Reg.*, Jap. *Hanadsuru, Hana-kadsura ;* a species of the preceding wtth creeping stems.

484. b. Illicium religiosum, *S. et Z.*, Jap. *Shikimi, Hanashiba ;* an evergreen tree of the order Magnoliaceæ growing wild in mountains of warm regions. It attains to a height of about 10 fts. In late spring it bears yellowish polypetalous flowers on the axils of the leaves and at the top of the young branches. The fruits ripen in autumn and produce seeds which contain a deadly poison. As the fruits have an aromatic flavour, they are exported to China and used instead of Illicium anisatum. The leaves have also a fragrant odour, and are used to odorn the vases offered to Buddha. An incense is prepared from the dried leaves by reducing to powder.

485. Macleya cordata, *R. Br.*, Jap. *Champa-giku, Takeni-gusa ;* a perennial herb of the order Papaveraceæ growing wild 5–6 fts. high. In autumn it bears small white flowers on the branches divided at the top of the stem, being succeeded with small pods. The leaves and stems contain a yellow juice. Bamboo becomes soft when boiled with this plant. The decoction of the stems and leaves is used to destroy injurious insects.

486. Chelidonium majus, *L.*, Jap. *Kusa-no-ō ;* a biennial herbaceous plant of the order Papaveraceæ growing wild everywhere. In spring it grows to a height of about 1 ft., bearing

4 petaled yellow flowers, which are succeeded with pods. This plant contains a yetlow juice.

487. Coriaria japonica, *A. Gr.*, Jap. *Doku-utsugi ;* a deciduous shrub of the order Coriariaceæ growing wild in bushes and on river banks. It is a diœcious or monæcious plant. It blooms in panicles, and the female flowers are succeeded with round red fruits, which are very pretty, but poisonous.

488. Rhus toxicodendron, *L.*, Jap. *Tsuta-urushi ;* a deciduous climber of the order Anacardiaceæ growing wild in forests and climbing on other trees. The leaves are ternate, and the flowers and fruits resemble those of R. vernicifera. The stem is used for dying.

489. Desmodium laburnifolium, *Dc.*, Jap. *Uji koroshi, Miso-naoshi, Miso-kusa ;* a deciduous shrub of the order Leguminosæ growing wild in warm provinces. It attains to a height of about 1 ft., but it is mostly herbaceous. In summer it shoots forth a panicle, and yields yellowish white papilionaceos flowers, being succeeded with long pods covered with hair. The leaves are used to kill the worms produced in *miso* (a kind of sauce).

490. Cicuta virosa, *L.*, Jap. *Doku-jeri. Ō-jeri ;* a triennial herb of the order Umbelliferæ growing wild in ponds and marshes. It resembles *Seri* (58) in form, but larger and poisonous. Early in spring, its stem is used as a pot-plant, being called *Chōmē-chiku.* The dried petioles are used for fastening as cord.

491. Andromeda japonica, *Th.*, Jap. *Asebo, Asebi, Asemi ;* an evergreen shrub of the order Ericaceæ, growing in mountains often 10 fts. high. Early in spring it produces bunches of campanulate small white drooping flowers. It is used as an ornamentel pot-plant. The leaves contain a violent poison, and the decoction is used to destroy injurious insects.

492. Buddleya curviflora, *Lindl.*, Jap. *Fuji-utsugi ;* a deciduous shrub of the order Loganiaceæ growing wild in mountains or on river banks, attaining to a height of 3-4 fts. The young branches are four sided and provided with alae. In summer it produces purplish pink flowers in panicles. The branches and leaves are used to intoxicate fishes for catching.

493. **Datura alba**, *Nees.*, Jap. *Chosen-asagao;* an annual solanaceous plant brought from Corea about the year 1744. It is sown in spring, and grows to a height of 3-4 fts. In autumn it bears white funnel formed flowers, with round seeds enclosed in a prickly capsule. The seeds, flowers, and leaves are poisonous.

494. **Solanum nigrum**, *L.*, Jap. *Inu-hōdsuki, Kuro-hōdsuki, Nasubi-sennari;* an annual herb of the order Solanaceæ growing wild everywhere. In summer it grows to a height of 2-3 fts., and bears 5 parted white flowers in clusters, being succeeded with round black fruits.

495. **Capsicum anomalum**, *Fir. et Sav.*, Jap. *Hadaka-hōdsuki, Tachi-hiyodori, Yama-hōdsuki;* an annual or sometimes biennial herbaceous plant of the order Solanaceæ growing wild in woods and bushes. It attains to a height of 20-30 fts., and resembles the preceding. The ripe berries are of a pretty pink colour.

496. **Scopelia japonica**, *Max.*, Jap. *Hashiridokoro;* a perennial herb of the order Solanaceæ growing wild in valleys. Early in spring, the young plant shoots dark purple leaves which turn green afterwards. It grows to a height of about 1½ fts., producing purple campanulate flowers in the axils of leaves, and then green round pea-sized berries.

497. **Solanum lyratum**, *Th.*, Jap. *Hiyodori-jōgo, Horoshi;* a perennial wild climber of the order Solanaceæ. In summer it shoots forth peduncles from the axils of leaves, bearing small white flowers, which are succeeded with small red round berries.

498. **Solanum Dulcamara**, *L.*, var. ovatum. *Dunal.*, Jap. *Maruba-no-hiyodorijōgo, Maruba-no-horoshi;* a close ally of the preceding. Its stem grows as a vine, but the plant is smaller. The flowers are light purple, and the berries red.

499. **Phytolacca acinosa,** *Roxb.*, var. esculenta, *Max.*, Jap. *Yamagobō ;* a perennial herb of the order Phytolaccaceæ growing wild, but also cultivated for its edible leaves. The stem attains to a height of 3-4 fts. In summer it produces panicles with small white flowers, which are succeeded with red berries.

500. **Clerodendron squamatum,** *Bahl.*, Jap. *Hi-giri*, *Tō-giri ;* a deciduous shrub of the order Verbenaceæ growing in warm regions, attaining to a height of 3-4 fts. From summer to autumn, it bears 5 petaled red flowers in bunches The flowers stretch out long stamens, and their calyx are also bright red. It is planted in gardens for ornamental purposes.

501. **Daphne kiusiana,** *Miq.*, Jap. *Koshō-no-ki ;* an evergreen shrub of the order Thymeleaceæ growing in shady places in mountains. Its shape resembles the Daphne odora (394), and it attains to a height of 3-4 fts. Early in spring it yields small yellowish white tubular flowers in clusters at the end of braches, being succeeded with oblong red berries. As the berries have the taste of pepper, the name *Koshō-no-ki* (pepper tree) is derived.

502. **Daphne pseudo-mezercum,** *A. Gray*, Jap. *Oni-shibari*, *Natsu-bōdsu*, *Sakura-kōzo ;* a deciduous shrub of the order Thymeleaceæ growing wild in bushes and on sea-coasts, attaining to a height of 3-4 fts. In spring it produces yellowish green flowers, and then red berries. The leaves fall in summer. The bast is strong, and used for manufacturing paper. A variety grown in the province of *Echigo* has large leaves and yellow fragrant flowers.

502. b. **Daphne genkwa,** *S. et Z.*, Jap. *Fuji-modoki*, *Chōji-sakura*, *Satsuma-fuji ;* a deciduous shrub of the order Thymeleaceæ, growing 3-4 fts. high. In spring it produces small purple tubular flowers in clusters before sprouting. It is planted in gardens on account of its pretty flowers.

503. **Euphorbia lathyris,** *L.*, Jap. *Horutosō ;* a bien-

nial herbaceous plant of the order Euphorbiaceæ brought by Portuguise about the year 1533. It attains to a height of 3-4 fts. In summer it bears flowers, which are succeeded with fruits about the size of a finger head.

504. **Euphorbia lasiocaula**, *Boiss.*, Jap. *Takatōdai ;* a perennial herb of the order Euphorbiaceæ growing wild in mountainous regions. It resembles the preceding in shape.

505. **Euphorbia sieboldiana**, *Morr.*, Jap. *Natsu-tōdai;* this resembles very much the preceding in shape, with shorter stems and broader leaves.

506. **Euphorbia helioscopia**, *L.*, Jap. *Tōdai-kusa ;* a small variety of the preceding, with short and creeping stems.

507. **Euphorbia palustris**, *L.*, Jap. *No-urushi ;* it resembles E. lasiocaula (504), growing in the form of a bush in watery places.

508. **Croomia japonica**, *Miq.*, Jap. *Nabe-wari, Kawanasubi ;* a perennial herb of the order Smilaceæ growing in shady places of mountains. The stem attains to a height of about 1 ft. Early in summer it shoots forth slender branched peduncles in the axils of leaves, bearing 4 petaled yellowish green flowers. The stem and leaves irritate the tongue.

509. **Nerine japonica**, *Miq.*, Jap. *Higan-bana, Shitamagari, Manju-shake ;* a bulbous plant of the order Amaryllideæ growing wild everywhere. In winter its leaves come forth and die in summer. In autumn the peduncles grow to a height of about 1 ft., and bear several flowers in clusters at the top. They are 6 petaled and of a deep red colour, having long stamen. Vulgar people eat the bulbs by drying and steaming. An inferior starch is obtained from them. The leaves and flowers of this and the next plant are produced at different times.

510. **Lycoris sanguinea**, *Maxim.*, Jap. *Kitsune-no-kamisori ;* a species resembling the preceding, growing wild in shady places.

The leaves are light green, and grow straight upward not crowded in one place. The flowers are orange red, and the roots have the same form and quality as the preceding. There is also a variety with white flowers.

511. Veratrum album, *L.*, var. grandiflora, Jap. *Baikeisō, Hai-no-doku;* a perennial herb of the order Meranthaceæ growing wild in moist places in mountains. The stalk is 3–4 fts. high, and its branches bear 6 petaled yellowish white flowers. Male and female flowers are separated on different plants, but sometimes complete flowers are found. The roots are poisonous, and are used to destroy flies and other injurious insects.

512. Veratrum stamineum, *Max.*, Jap. *Kobaikei, Shishi-no-habaki;* a species of the preceding of a smaller size. Its quality and use are the same, but its flowers do not have green veins as the preceding.

513. Veratrum nigrum, *L.*, Jap. *Shuro-sō, Nikkō-ran;* a close ally of 511 with narrower leaves. The flowers are dark purple, and have a disagreeable odour. The roots and young sprouts are covered with something like chamaerops fibres, whence the name *Shuro-sō* (chamaerops herb). A small sized species is called *Ao-yagi-sō,* which is the same in quality and use.

514. Alisma plantago, *L.*, Jap. *Saji-omodaka;* a perennial herb of the order Plantagineæ growing wild in swampy places. In summer its stem grows to a height of 2–3 fts. and is divided into several branches, bearing 3 petaled small white purple-shaded flowers. The tuberous roots are dried and used as medicine, but the stems and leaves are poisonous. There is another species with narrow leaves called *Hera-omodaka.*

515. Arisaema præcox, *Deverise.*, Jap. *Yuki-mochi-sō;* a bulbous herb of the order Aroideæ growing wild in mountains. The leaves are ternate, and the stem attains to a height of about 1½ fts. It bears spathaceous flowers with snow white pistils.

516. Arisaema thunbergii, *Blume.*, Jap. *Maidsuru-*

tennanshō; a bulbous plant of the order Aroideæ growing wild in shady places in mountains. The leaves are parted into 4 on the petiole. It grows to a height of about 1 ft., and bears flowers in spathes. The head of the pistils is small and pointed.

517. Arisæma serratum, *Th.,* Jap. *Hebi-no-daihachi, Mamushi-gusa;* it resembles the preceding in quality. The leaves are broad and serrated. The stem is covered with a purplish brown variegation like the colour of a snake, whence the name is derived.

518. Arisæma ringens, *Schott.,* Jap, *Musashi-abumi;* this resembles Arisaema praecox (515). The flowers have the form of a stirrup, whence the name.

519. Pinellia tuberifera, *Ten.,* Jap. *Karasu-bishaku, Hesobe;* a bulbous plant of the order Aroideæ growing abundantly in fields. It is a small weed, but is injurious to other cultivated plants. The stem grows a height of 7-8 inches, bearing small long dark purple spathaceous flowers. The roots are used as medicine when dried. There is a kind, with the stem about 1 ft. high, and larger leaves, flowers, and roots, called *Ōba-hange.*

520. Arisæma japonicum, *Bl.,* Jap. *Tennanshō, Yabukonniyaku;* a bulbous plant of the order Aroideæ growing wild in shady places, being the most common of this family. It resembles A. serratum (517) in shape, but the leaves have several divisions. The stems and leaves are quite green, but the spathe is stripped with purple and provided with a oblong finger-sized pistil, being followed with small round berries. The roots are dried and used for medicine, and also to destroy injurious insects.

521. Arisæma thunbergii, *Blume.,* var. foliolis angustioribus, Jap. *Urashima-sō;* it resembles A. thunbergii (516), but the tips of the pistils are slender drooping in the form of a fishing line. The roots are tuberous.

522. Jap. *Tengu-no-karakasa;* a terrestial fungus growing under trees in autumn. It grows to a height of about 6 inches. The size of the pilius is about 5 inches in diameter. The stem is provided with something like the guard on the hilt of a sword. This and the following fungi are all poisonous.

523. Jap. *Tōjin-take;* a terrestial fungus growing in forests in autumn or summer. It attains to a height of 4-5 inches. Its thallus is about 5 inches in diameter, and yellow in colour. The gills are pink.

524. Jap. *Ochiba-take;* a terrestial fungus appearing in fallen leaves under woods in late autumn.

525. Jap. *Moyegi-take;* a terrestial fungus appearing in moist shady places under woods late in autumn. The shape and colour resemble very much those of *Hatsu-dake* (143), but with quite a different nature.

526. Jap. *Usu-take;* a terrestial fungus growing in shady places under woods late in autumn. The middle part of the thallus is concaved like a mortar, whence the Japanese name is derived. Both the thallus and stype are white.

527. Jap. *Haikoroshi-take;* a terrestial fungus growing in shady places under woods in late autumn. When it comes up at first it is like an egg, but bursts afterwards. It has a height of about 4 inches, and is white and gummy.

528. Jap. *Komusō-take;* a terrestial fungus growing in shady places under woods from autumn to winter. Its thallus forms a thin lining like a net over the top. Its surface is covered with yellow dust and has a disagreeable odour as a rotten animal.

529. Jap. *Haitori-take;* a terrestial fungus growing in shady places in late autumn. It resembles *Shimeji-take* (142) in shape. The thallus is greenish white. It is used to poison flies by mixing it into boiled rice.

530. Jap. *Tengu-take;* a terrestial fungus growing in moist shady places in the beginning of winter. It resembles very much *Shimeji* (142) in shape and colour, but about twice in height.

531. Jap. *Hotaru-take, Tsukiyo-take;* a terrestial fungus produced under grasses in plains. It resembles *Shimeji* (142) in shape. In night this fungus gives a phosphoric light, whence the Japanese name is derived. There are several sorts with the same quality and name.

532. Jap. *Momiji-take, Warai-take;* a parasitic fungus growing on maple trees, resembling *Matsu-take* (mush-room) in shape. If a man eats this fungus, he will be poisoned and wil laugh, whence the name is derived.

CHAPTER XXI.—TIMBER TREES AND BAMBOOS.

This Chapter contains timbers and bamboos, the stems of which are used for various purposes, as the buildings of palaces, houses, bridges, ships, railways, telegraphs,etc. Though they are different in qualities, as hard or soft, and flexible or brittle, yet all of them have their respective uses. They are also used as fuel.

533. Magnolia hypoleuca, *S.* et *Z.*, Jap. *Hō-no-ki;* a deciduous tree of the order Magnoliaceæ growing wild in mountains, attaining to a height of 40-50 fts. After the leaves shoot forth, it opens pale yellow flowers, which odour is too strong. The fruits are oval-shaped, and expose many red berries. The wood is yellowish or greenish, and is very fine in structure, being suitable to make tailor's tables, stamp-blocks, and many other things. The charcoal prepared from this wood is much prized by lacquer-makers and gold-smiths for polishing.

534. Cercidiphyllum japonicum, *S.* et *Z.*, Jap. *Katsura;* a deciduous tree of the order Magnoliaceæ growing in mountains 40-50 fts. high. The two sexes of flowers grow separately on different plants. In spring, it produces pink flowers, before it sprouts, and in autumn it produces small pods. The wood is brown and fine grained, and is used for making chess-boards, tables, boxes, Japanese wood-shoes, and many other articles.

535. **Æsculus turbinata,** *Blume.*, Jap. *Tochi-no-ki;* a deciduous tree of the order Sapindaceæ growing in mountains 40-50 fts. high. In early summer, it produces white pink-shaded flowers in panicles on the branches. In late autumn, its round fruits ripen and expose nuts about 1 inch in size. The nuts are dried and eaten. The wood is pale yellow, and resembles the preceding two in use. It is manufactured by turners for making trays, plates, bowls, etc. The old wood is variegated and very pretty.

535. b. **Acer pictum,** *Th.*, Jap. *Itaya-momiji*, *Tokiwa-kayede;* a deciduous tree of the order Aceraceæ, growing principally in northern countries. Its stem attains to a height of 40–50 fts. The wood is light brown and fine grained. As the grains of the old wood are very beautiful with a circular figure, it is used for ornaments of rooms or to make boxes. Sugar is obtained from the juice of the fresh stems.

536. **Melia japonica,** *G. Don.*, Jap. *Sendan;* a deciduous tree of the order Meliaceæ growing to a height of 20–30 fts. In summer, its purple flowers open in panicles, being succeeded with oval fruits about 1 inch long. The fruits ripen in winter und become yellow. The wood resembles Zelkowa *keaki* (557). Especially the old wood is very beautiful. Besides this, there is a species called $\bar{O}sendan$, which grows more rapidly and largely.

. 537. **Cedrela chinensis,** *A. Juss.*, Jap. *Chanchin;* a deciduous tree of the order Cedrelaceæ growing to a height of 30–40 fts. In summer its small white flowers open in panicles and are succeeded with pods. When the pods are fully ripe in autumn they expose seeds which fly away. The leaves are disagreeably odorous, especially at night. A variety with red young leaves is called *Akebono*. The wood is red and hard. Its use is nearly the same as the former.

538. Evodia glauca, *Miq.*, Jap. *Kihada;* a deciduous tree of the order Zanthoxylaceæ growing in mountains 30-40 fts. high. It is a diaecious plant. In summer it yields fine yellow flowers, being succeeded with round black fruits, which are called *Shiko-no-hei* and used as medicine. The bark is used as a yellow dye and also for medicine. The wood is hard and used to make boxes and many other furnitures.

539. Evonymus europænus, var. hamilitonianus, Jap. *Yama-nishikigi;* a deciduous shrub of the order Celastraceæ growing wild to a height of about 10 fts. Its light green flowers open at the same time with its leaves. When the fruits are ripe, their outer skins burst and expose red seeds. As the wood is pale yellow and fine grained it is used for making combs, ax-handles, and many other articles. It is also used by turners.

540. Hovenia dulcis, *Th.*, Jap. *Kenpo-nashi;* a deciduous tree of the order Rhamnaceæ growing wild to a height of several fts. In summer it yields small white flowers, and produces small fruits. The peduncles are delicious when they become fleshy and turn purple brown. The wood is hard, and especially when old it is very pretty. It is used to make tables, boxes, and other articles of furniture.

541. Sophora japonica, *L.*, Jap. *Yenju;* a deciduous tree of the order Leguminoceæ planted in gardens, growing to a height of 20-30 fts. In summer it bears pale yellow papilionaceous flowers, being succeeded with pods containing many seeds. The pod is contricted between the seeds like a rosary. The wood is fine grained and hard, being used for many purposes. *Inu-yenju* (Cladrastis amurensis) grows wild, and its wood is employed by turners. As the wood is black at its centre, it is also called *Kuro-yenju* (black sophora).

542. Prunus pseud-cerasus, *Lindl.*, Jap. *Sakura;* a deciduous tree of the order Rosaceæ growing wild in mountains, being called *Yama-zakura* (mountain-cherry). The one pitured in this book is called *Some-yoshino*. It attains to a height of

20-80 fts. In spring, it opens light pink flowers, succeeded with small red purple fruits which have a subacid taste. The wood is brown and fine grained, being used for engraving and many other purposes. The bark is strong and smooth, being suited for knitting, fastening, etc.

543. Prunus groyana, *Max.*, Jap. *Uwamidsu-sakura;* a deciduous tree of the order Rosaceæ growing wild about 10 fts. high. In early summer it yields small white flowers disposed in panicles, being succeeded with pea-sized fruits, which are eaten by salting. The wood is yellowish red and fine grained. It is used in the same way as the preceding.

544. Distylium racemosum, *S. et Z.*, Jap. *Isu, Hyon-no-ki;* an evergreen tree of the order Hamamelideæ much growing wild in warm regions, and also planted in cold countries. It attains to a height of 20-30 fts. After the new leaves shoot forth, it opens small dark red flowers in chusters. Insects very often make their nests on the leavesl, and afterwards the nests become empty shells, whichare called *Hiyon-ko*. The wood is hard and fine grained, and its colour is red, with a dark brown centre. It is used to make combs, eating sticks, musical instruments, boxes, and other ornamental works. The ash of this wood is used as a glazing material of porcelain.

545. Acanthopanax ricinifolium, *S. et Z.*, Jap. *Hari-giri, Sen-no-ki, Yama-giri;* a wild deciduous tree of the order Araliaceæ growing to a height of 20-30 fts. The young trees are very thorny, but when old the thorns fall off. In summer it yields small yellowish white flowers in panicles. The wood is used for boxes and turnery. The young leaves are eaten.

546. Cornus macrophylla, *Wall.*, Jap. *Midsuki, Midsu-no-ki, Midsu-kusa;* a deciduous tree of the order Cornaceæ found everywhere growing to a height of about 20 fts. In early summer it bears small white flowers arranged in an umbel, being succeeded with small round dark purple fruits. The wood is white and fine grained, but soft. It is used for turnery.

547. Styrax japonicum, *S.* et *Z.*, Jap. *Yego-no-ki, Rokuro-gi, Chisha-no-ki;* a wild deciduous tree of the order Styracaceæ growing to a height of about 10 fts. In early summer it droops peduncles from the axiles of leaves, and opens white flowers, which are succeeded with small round fruits containing hard seeds. An oil is made from the seeds. Its wood is white and fine grained, being used mostly for the handles of umbrellas.

547. b. Deutzia scabra, *Th.*, Jap. *Utsugi;* the wood of this tree is white and fine grained, being used for mosaic-works, and wooden nails.

548. Viburnum opulus, *L.*, Jap. *Kanboku;* a deciduous shrub of the order Caprifoliaceæ growing wild in cold regions, attaining to a height of 8–9 fts. In early summer it bears white flowers in an umbel being succeeded with small red fruits. As the wood is white, fine-grained, and flexible, it is used mostly for tooth-brushes.

549. Diospyros kaki, *L.*, Jap. *Kaki;* as this plant is described in the Chapters of dye plants (368) and fruit trees (194), it is only mentioned here of its timber. The one, which central part of the wood is black and hard, is called *Kuro-kaki* (black ebony), and the one with black stripes is called *Shima-kaki* (striped ebony). Their qualities are not inferior to *Kokutan* wood. They are used for turnery, mosaic works, and many other articles. The one pictured in this book is what we call *Yama-gaki* (mountain persimmons). The wood of *Shinano-gaki* (368) is also good.

550. Fraxinus mandshurica, *Rupr.*, Jap. *Shioji, Yachi-damo;* a deciduous tree of the order Oleaceæ growing wild in cold regions, attaining to a height of 20–30 fts. In early summer it yields small narrow petaled flowers in clusters, being succeeded with small pods. The wood is yellowish white, hard, and fine grained, being used for scale-rods, spokes of wheels, handles of several articles, and many others.

551. Olea aquifolium, *S.* et *Z.*, Jap. *Hiiragi;* an evergreen tree of the order Oleaceæ growing wild, and also planted in gardens. The stem attains to a height of 10 fts. In autumn it bears fragrant small white flowers in clusters in the axils of leaves, being succeeded with small oval fruits which are purplish blue when ripe. The wood is white, hard, and fine grained, and is used to make combs, chop-sticks, engravings, abacus, wooden toys, chess-men, etc.

552. Paulownia imperialis, *S.* et *Z.*, Jap. *Kiri;* a deciduous tree of the order Scrophulariaceæ planted everywhere growing to a height of 20–30 fts. Before sprouting it bears purple or white labiate flowers in panicles. The fruits have capsules, shaped like a pigeon's egg, and expose many small winged seeds. The wood is soft and white, and hollow in the centre. The fine grained old wood is called *Shima-giri*, and is used for making musical instruments, various cases, tables, etc.

553. Cinnamomum camphora, *Nees.*, Jap. *Kusu-no-ki;* an everygreen tree of the order Lauraceæ produced in warm regions, growing 30–40 fts. high and several feet in circumference. In early summer, it produces long peduncles from the axils of leaves, and bears small pale yellow flowers. The fruits are pea-sized and black. The wood is gray and fine grained, and when old it becomes harder and brown. The old wood has a circular figure and cloud-like variegation. It is used for building houses and ships, and also to make book-cases, garment-cases, and many other articles, but it is not suitable for table-vessels, because it is too odorous. Camphor is made from this wood.

553. b. Machilus thunbergii, *S.* et *Z.*, Jap. *Tama-kusu, Ao-kusu, Yama-kusu, Kara-damo, Inu-kusu;* an evergreen tree of the order Lauraceæ growing in warm regions, attaining to a height of several feet. In autumn it blooms, and in the following summer purplish black fruits are produced. The wood is dark brown, hard, and fine grained, and the old wood has beautiful whirls and cloud-like variegation. An oil is taken from the seeds, and the bark is used for dying.

554. Lindera hypoleuca, *Max.*, Jap. *Kuromoji ;* a deciduous shrub of the order Lauraceæ, growing wild in mountains of many countries, attaining to a height of 7-8 fts. It bears small yellow flowers in clusters on the branches, being succeeded with pea-sized round fruits which are black when ripe. The bark is black and fragrant. The wood is white and fine grained, and is used for making tooth-picks. The branches and stems are used for fences.

555. Rottlera japonica, *S.* et *Z.*, Jap. *Akame-gashiwa ;* a deciduous tree of the order Euphorbiaceæ growing wild in many regions, attaining to a height of 20-30 fts. It is a diæcious plant. In summer it bears small pale yellow flowers in panicles, being succeeded with many thorny fruits which burst and expose small black seeds when they ripe. As the wood is red and fine grained, it is used for boxes and pillars of Japanese houses.

556. Buxus japonica, *Mull.*, Jap. *Tsuge ;* an evergreen shrub of the order Euphorbiaceæ, growing wild in mountains of warm regions. The famous places for this plant are *Mikura-jima* of *Idsu*, *Asakuma-yama* of *Ise*, *Kosho-yama* of *Chikuzen*, the provinces of *Satsuma* and *Ōsumi*, *Okinawa*-island, etc. The shape of the leaves differs according to the places where they grow, and the colour of the wood also differs. It attains to a height of about 10 fts. It is a monœcious plant. In summer it bears yellowish flowers, being succeeded with pea-sized fruits. The wood is yellow and fine grained, and in hardness it is superior to many other woods. It is very valued to make combs, engravings, stamps, etc.

557. Zelkowa keaki, *Sieb.*, Jap. *Keyaki ;* a deciduous tree of the order Urticaceæ growing wild or planted everywhere, attaining to a height of 30-60 fts. In spring it produces male and female flowers separately at the same time with new leaves. The flowers are small and yellow, and are succeeded with small flat seeds. The young wood is yellowish white, hard and tough, being used for houses, ships, tables, boxes, handles of various

— 139 —

articles, etc. The old wood is dark brown and hard, and is prized for the whirling and cloud-like variegation of its grain, being used for various articles, though it is rather brittle.

558. Celtis sinensis, *Pers.*, Jap. *Ye-no-ki ;* a deciduous tree of the order Urticaceæ growing wild everywhere, attaining to a height of 40-50 fts. It has diaecious, monaecious or perfect flowers. The flowers are small and yellow, being succeeded with small round fruits which turn red when fully ripe. The sweet pulp of the fruits is edible. The wood is yellowish white and fine grained, but soft, and it is used for turnery.

558. b. Morus alba, *L.*, Jap. *Kuwa ;* as the wood of this tree (294) is yellow and hard, it is valuable for making various vessels.

559. Ulmus parvifolia, *Sacq.*, Jap. *Aki-nire, Ko-nire ;* a deciduous shrub of the order Urticaceæ growing wild everywhere, attaining to a height of 20-30 fts. In summer it bears small light green flowers, being succeeded with flat pods. The wood is brown, hard and fine grained, being used for many purposes. A species called *Haru-nire* grows very fast.

560. Salix multinervis, *Fr.* et *Sav.*, Jap. *Kori-yanagi, Kobu-yanagi ;* a deciduous shrub of the order Amentaceæ growing wild near water. Those which grow in Province *Tajima* are esteemed as the best. The stems grow to a height of 6-7 fts. in groups. The stems are used for plaiting after the bark is taken off and bleached.

561. Populus tremula, *L.*, var. villosa, *Wesm.*, Jap. *Yama-narashi, Hako-yanagi, Maruba-yanagi ;* a deciduous tree of the order Amentaceæ growing wild in mountains in many districts. It is a diaecious plant. The stem grows to a height of 20-30 fts. In spring it produces catkins before it sprouts. The wood is white, fine grained and tough. It is used to make toothbrushes, boxes, engravings, etc.

562. Populus suaveolens, *Fisch.*, Jap. *Dero-yanagi*,

Doro-yanagi, Wata-no-ki; a deciduous tree of the order Amentaceæ growing in mountains of cold regions, attaining to a height of 20-30 fts. It is a diaccious plant, and produces catkins before the leaves come forth. When the seeds ripen, a kind of fibre like cotton is exposed. The wood is white, fine grained, soft and brittle, being principally used for matches.

563. Quercus glandulifera, *Bl.*, Jap. *Nara-no-ki, Konara;* a deciduous tree of the order Amentaceæ growing wild to a height of 20-30 fts. It is a monaecious plant. Before sprouting it produces catkins, with acorns on the cups. The wood is hard and strong, and is suited for fuel. The stems and branches are used to cultivate a kind of mush-room called *Shiitake* (140) on them. The kernels are eaten. The shape, quality and use of these species are almost common to each other.

563. b. Quercus crispula, *Bl.*, Jap. *Ōnara, Midsunara;* a species closely allied to the preceding. It is a large tree found wild in mountains. It has the same uses as before.

563. c. Quercus variabilis, *Bl.*, Jap. *Wata-nara, Watakunugi, Abemaki;* a species of Quercus (295) with thick bark which is used as cork. The new bark, grown after the outer bark was stripped off, is very suitable for this purpose. It requirse about 10 years to get good bark.

563. d. Quercus serrata, *Th.*, Jap. *Kunugi;* the wood of this tree (295) is the best as fuel.

564. Quercus acuta, *Th.*, Jap. *Aka-gashi, Ō-gashi;* an evergreen tree of the order Amentaceæ produced in warm regions, growing to a height of 20-30 fts. It is a monaecious plant. Its acorn resembles that of Q. glandulifera (563). The wood is red and hard, and is used to make wheels aud other articles.

564. b. Quercus gilva, *Bl.*, Jap. *Ichii-gashi;* an evergreen tree of the order Amentaceæ. As it produces edible acorns, it is described in the chapter of fruits (226. b.). It grows to a height of 30-60 fts. It is a monaecious plant. The wood is red and very strong, being valued for oars. This wood is commonly called also *Akagashi* as the preceding.

565. Quercus glauca, *Th.*, Jap. *Shira-kashi;* an evergreen tree of the order Amentaceæ produced in warm regions, growing to a height of 20-30 fts. It resembles Q. acuta (564), but the leaves are thinner and the acorns smaller. The wood is white, but the use is almost the same as Q. acuta.

566. Quercus phyllireoides, *A. Gray*, Jap. *Ubamegashi*, *Imame-gashi;* an evergreen tree of the order Amentaceæ produced in warm regions, growing to a height of about 10 fts. It resembles other oaks in shape, but as it grows very slowly, it is difficult to become a large tree. The wood is red and very hard. It is principally used for making oars, and also much used to make charcoal called *Bincho*, which is prized for its great heat.

567. Fagus silvatica, *L.*, Jap. *Buna-no-ki;* a deciduous tree of the order Amentaceæ produced in mountains of northern regions, growing to a height of 30-60 fts. It is a monaecious plant. It produces hairy fruits, which expose triangular kernels eatable by grilling. Oil may be taken from them. The bark contains tannin, and is used next to the oak. The wood is strong, being used for wooden spoons and turnery.

567. b. Castanea vulgaris, *Lamk.*, var. japonica, *D.C.*, Jap. *Kuri-no-ki;* the wood of this tree (221) is hard and durable, being suitable to be used in damp places.

567. c. Carpinus luxiflora, *Bl.*, Jap. *Soro-no-ki, Inushide;* a wild deciduous tree of the order Amentaceæ growing to a height of 20-30 fts. Barren and fertile flowers grow separately, and they are succeeded with drooping scaly cones or catkins. The wood is used to produce *Shii-take* or as fuel. The stem has an uneven surface, being used as pillars for curiosity.

568. Betula alba, *L.*, Jap. *Shira-kaba;* a deciduous tree of the order Amentaceæ growing wild in northern countries, attaining to a height of 30-40 fts. It is a monaecious plant, blooming in summer. The male flowers droop in catkins, and the female form a round scaly cone containing many small seeds between the scales. The bark of the wood is thin and easily peeled off, being used for plaiting and tying, and also to make various articles. The wood is white and fine grained, being used for boxes and turnery.

569. Betula alba, *L.*, var. tauschii, *Reg.*, Jap. *Mine-hari*, *Ono-ore ;* a deciduous tree of the order Amentaceæ growing in mountains of northern countries, attaining to a height of 20-30 fts. In summer it bears monaecious flowers, and are succeeded with small fruits like those of *Han-no-ki*. The wood is light brown, hard and fine grained, being used for making the reed of looms, combs, etc.

570. Pterocarya rhoifolia, *S.* et *Z.*, Jap. *Sawa-gurumi*, *Kawa-gurumi*, *Yasu-no-ki ;* a deciduous tree of the order Juglandaceæ, growing wild in mountains of cold regions, attaining to a height of 20-30 fts. The flowers are monaecious, and are succeeded with small fruits provided with wings. The wood is white, fine grained and light, and it is used for making boxes, Japanese clogs, etc. The polished bark is made into various articles, being called *Jukō-hi* in *Nikko*.

571. Juniperus rigida, *S.* et *Z.*, Jap. *Nedsumi-sashi*, *Muro ;* an evergreen tree of the order Coniferæ growing wild in mountains. The stem grows straight to a height of 20-30 fts. It is a diaecious plant. In summer it produces small flowers from the axils of leaves, being succeeded with round pea-sized black oily fruits. The wood is hard and yellow with a fragrant resinous odour, being used for ornamental pillars, shelves, and small articles.

572. Juniperus chinensis, *L.*, Jap. *Beni-byakushin*, *Ibuki ;* an evergreen tree of the order Coniferæ growing wild. The one which grows on a high mountain seems like a shrub, and the one which grows in plain stands straight attaining to a height of about 10 fts. In summer it produces monaecious flowers, and then fruits. The leaves have two forms. The wood is reddish brown, hard and fine grained with a fragrant resinous odour. The use is nearly the same as the preceding.

573. Thuja dolabrata, *L.*, Jap. *Hiba, Asunaro, Hinoki, Asuhi;* an evergreen tree of the order Coniferæ growing in mountains of northern provinces, attaining to a height of 30–40 fts. In summer it produces monaecious flowers and small balls. The wood is pale yellow, fine grained and lustrous, and is used for house-buildings and furnitures, being important next to *Hinoki.* The bark is used as *Maki-hada,* and also made into a rope-match.

574. Thuya japonica, *Max.*, Jap. *Nezuko, Gorō-hiba, Kurobe-sugi;* an evergreen tree of the order Coniferæ growing wild in mountains to a height of 20–30 fts. In summer it bears monaecious flowers. It resembles very much the preceding, but smaller and better as a garden plant. The wood is dark brown, resembling Cryptomeria japonica, and is made into tables and several other ornamental furnitures. There is a different sort called *Hime-asunaro,* which is fine and slender.

575. Thuya obtusa, *Benth* et *Hook,* Jap. *Hinoki;* an evergreen tree of the order Coniferæ growing in mountains. Those produced in the *Kiso* mountain in Province *Shinano* are very famous. It attains to a height of 30–60 fts., and in summer it produces monaecious flowers and then small balls. The wood is yellowish white, fine grained, and lustrous. This is one of the most useful timbers for house-buildings, bridges, etc., being very resistible against bending or contracting.

576. Chamaecyparis pisifera, *S.* et *Z.*, Jap. *Sawara;* an evergreen tree of the order Coniferæ produced almost in the same districts as the preceding. The shape is also nearly same, but the leaves are more pointed and the cone smaller. The wood is more yellowish and softer. It is used in the same way, but inferior to the preceding.

577. Cryptomeria japonica, *Don.*, Jap. *Sugi;* an evergreen tree of the order Coniferæ produced abundantly everywhere, growing to a height of 30-60 fts. It is one of the plants growing to a considerable height. In summer it bears monaccious flowers, producing cones about the size of a finger. The wood is fine-grained and light, being yellowish white on the outside and reddish brown inside. It is used for house and ship-buildings, bridges, boxes, tubs, and many other articles. The wood and bark are used to cover roofs, and the leaves are made into incence-sticks. The old wood when burried in a pond or marsh becomes dark green, and is very esteemed by the name of *Jindai-sugi*.

577. b. Cryptomeria japonica, S. et Z., Jap. *Yaku-sugi;* a variety of the preceding produced in the Island of *Yaku-shima* of Province *Ōsumi*. The old wood is brown, resinous, fine grained and hard. It is useful for making boxes, tables, and other ornamental furnitures.

578. Podocarpus chinensis, *Wall.*, Jap. *Maki, Inu-maki, Hitotsuba;* an evergreen tree of the order Coniferæ found in mountains of many provinces, attaining to a height of 20-30 fts. It is a diaecious plant. The male flowers form drooping catkins, and the female produce berries formed of two pieces, the under one of which is red, freshy and edible, and the upper one is a green and pea-sized seed. The wood is white and fine grained, and is valuable for building. The wood of *Kōya-maki* (707) is also called *Maki*-wood.

578. b. Sciadopytis verticillata, S. et Z., Jap. *Kōya-maki, Kusa-maki;* the wood of this tree (707) is esteemed for its durability against moisture. The bast of the trunk is called *Maki-hada*, and is used to stop the leaking of water.

578. c. Torreya nucifera, S. et Z., Jap. *Kaya;* the wood of this conifer (217) is yellowish white and fine grained, with a fragrant resinous odour, and is used for various buildings in moist places. It is also used to make chess-boards, chess-men, abacus, etc.

579. Podocarpus nageia, *R. Br.*, Jap. *Nagi;* an evergreen tree of the order Coniferæ produced in warm regions, attaining to a height of 30-60 fts. It is a diaecious plant. The male flowers produce yellowish drooping catkins, and the female yield round fruits about the size of a finger. The wood is white and fine grained, and is used for furnitures and house building.

580. Toxus cuspidata, *S.* et *Z.*, Jap. *Ichii, Araragi, Onko;* an evergreen tree of the order Coniferæ growing in mountains of various provinces and especially in *Zezo.* It attains to a height of 20-30 fts. It is a diaecious plant. The male flowers droop greenish brown catkins, and the female produce small round fruits which are red and sweet when ripe, containing black seeds. The wood is brown and fine grained, with a fragrant odour. It is very good and highly prized to make tables, boxes, and many other articles.

581. Pinus thunbergii, *Parlat.*, Jap. *Kuro-matsu, O-matsu;* an evergreen tree of the order Coniferæ growing plentifully on the sea coasts of southern provinces, attaining to a height of 30-60 fts. It is a monaecious plant. The male flowers form small catkins, and the female produce small cones which grow to the size of a small wrist in next autumn and then the scoles burst to scatter about the winged seeds. The wood is reddish white, fine-grained, and very resinous. It is used to build houses, ships, bridges, etc.

582. Pinus densiflora, *S.* et *Z.*, Jap. *Aka-matsu, Me-matsu;* an evergreen tree of the order Coniferæ growing wild abundantly, attaining to a height of 30-60 fts. It is a diaecious plant. It is closely allied to the preceding, but the leaves are softer and the cones smaller. Its use and nature are also nearly the same. Its round stem with bark is used as pillars and ornaments in the rooms of Japanese houses.

583. Pinus parviflora, *S.* et *Z.*, Jap. *Hime-ko-matsu;* an evergreen tree of the order Coniferæ growing wild in mountainous districts of many provinces, attaining to a height of 30-60 fts. It resembles the preceding in form, but it has 5 needles in leaves instead of 2. The wood is reddish white, fine-grained and resinous. It use is nearly the same as *Kuro-matsu*, but superior.

584. Larix leptolepis, *Gord.*, Jap. *Kara-matsu*, *Fuji-matsu*; a deciduous tree of the order Coniferæ growing wild in mountains, attaining to a height of 30–50 fts. It bears monaecious flowers, and the fertile flowers yield cones of the size of a thumb. The wood is hard and reddish brown, being used for buildings and esteemed for its durability.

585. Abies firma, *S.* et *Z.*, Jap. *Momi*; an evergreen tree of the order Coniferæ growing wild everywhere, attaining to a height of 30–60 fts. In summer it produces monaecious flowers, female yielding cones about 4–5 inches long. The wood is soft and white, and is used for buildings or for making boxes.

586. Tsuga sieboldii, *Carr.*, Jap. *Tsuga*, *Toga*; an evergreen tree of the order Coniferæ growing wild in mountains in many provinces, attaining to a height of 30–40 fts. In summer it produces barren and fertile flowers, and yields small cones about the size of a thumb. The wood is hard and reddish white, being used for buildings and many other articles. The bark is used to dye fishing nets.

587. Abies mengiesii, *Lond.*, Jap. *Tōhi*; an evergreen tree of the order Coniferæ growing in mountains of many provinces, attaining to a height of 30–60 fts. In summer it produces barren and fertile flowers, being succeeded with cones resembling those of Abies firma, but slender. The wood is white, fine-grained and flexible being used to make round boxes and many other articles. There are several sorts of this genera, such as *Shinko-matsu*, *Matsu-hada*, etc., which are nearly the same in use and shape.

588. Abies veitchii, *Henk.* et *Hochst.*, Jap. *Shirabe*, *Shirabi*, *Shirabio*; an evergreen tree of the order Coniferæ growing wild in mountains of many districts, attaining to a height of 30–60 fts. In summer it produces barren and fertile flowers. Its cones resemble those of Abies firma. The wood is white and soft, being used for nearly the same purposes as the preceding. The *Yezo-matsu* of *Hokkaidō* is closely allied to this.

589. Phyllostachys Quilioi, Riv. Jap. *Madake, Kawa-take ;* an evergreen bamboo of the order Gramineæ planted every where, growing wild in warm regions. It grows 60-70 fts. high and about 1½ fts. in circumference. In May and June its young sprouts are eaten as vegetables. The sheathes covering the young bamboo have many uses. The best season to cut the stems is from the middle autumn to the middle winter. The stems are used for buildings and many purposes.

590. Phyllostachys mitis, Riv.   Jap. *Mōsō-chiku ;* a bamboo planted mostly in warm regions for the sake of its young sprouts, which are eaten as a vegetable (125. b.) It attains to a height of 40-50 fts., the circumference of its stem being about 2 fts. It is inferior to the preceding in quality, but as it is larger it is used for buildings and to make several sorts of vessels. The sheath is used next to the preceding.

591. Arundinaria japonica, *S.* et *Z.*, Jap. *Me-dake, Nayo-take, Shinobe-take ;* a bamboo growing mostly in warm regions, being used especially for a shelter on sea-shores. The stem is slender, 20-30 fts. high, and 5-6 inches. in circumference. Autumn or winter is the best time for cutting it. It is used for buildings, hedges, fences, handles, rods, Japanese fans, and many other articles.

592. Bambusa puberula, *Miq.*, Jap. *Ha-chiku, Kure-take ;* a bamboo planted everywhere much growing wild in mountains of warm regions. The greatest stem is 40-50 fts. high and 2 fts. in circumference. It resembles *Madake* (589) in quality, form, and use. It is admired on account of its spotless sheath. The stems are used as ropes, and the roots are also used as sticks and whips. **Phyllostachys nigra,** *Munro.*

593. ＿＿＿＿＿ ＿＿ , ＿＿＿＿＿ ., Jap. *Goma-dake, Kuro-chiku ;* a bamboo resembling the preceding in form, with thin leaves. The greatest stem is 20 fts. high, having a circumference of 5-6 inches. That with a black-spotted stem is called *Gomatake*, and that with a black stem *Kuro-chiku*. It is used to make sticks, handles of several articles, tables, book stands, and other various kinds of furnitures.

594. Bambusa senanensis, *Fr.* et *Sav.*, Jap. *Sudsutake ;* a bamboo growing wild in mountains, attaining to a height of 5-6 fts., with broad pointed leaves 5-7 inches long. The stem is slender, but strong, being used to make baskets and mats by splitting. It sometimes yields fruits which are used as food.

594. b. Bambusa, Jap. *Nemagari-take, Magari-take, Jintake ;* a bamboo growing in northern provinces. It resembles the preceding in form and quality, its stem being bent near the roots.

595. Bambusa chino, *Fr.* et *Sav.*, Jap. *Hakone-dake, No-dake ;* a celebrated bamboo of *Hakone* mountains in Province *Sagami*. It grows to a height of about 10 fts. It is used for making hedges and also to make Japanese pipes, brush handles, fans, baskets, ropes etc.

Note.—Though the timbers and bamboos above mentioned are principally used for buildings, furnitures, or fuel, yet some of them yield edible fruits, some are employed for various other purposes, as paper-making, fastening, etc., and some are also planted as ornaments in gardens and avenues. Generally speaking, if we refer to the uses of woods, there is no plant whatever that has not a certain use. Even a small shrub and a tiny bamboo may be used as handles of various articles, sticks, whips, etc. There are also many other plants which stems are used as woods ; for examples Pyrus ussuriensis (189), Zizyphus vulgaris (188), Photinia japonica (192), and Juglans sieboldiana (219) in the chapter of fruit trees ; Camellia japonica, Elaococca condata (313), Rhus succedanea (320), and Rhus vernicifera (321) in the chapter of oil and wax plants ; Chamærops excelsa (711), Rhapis flabelliformis (712), and Bamboos in the chapter of evergreen gardentrees and shrubs. The various trees found in southern islands are omitted here.

— 149 —

CHAPTER XXII.—DECIDUOUS GARDEN-TREES AND SHRUBS.

This chapter contains the ornamental plants, which leaves fall in autumn. They are planted in gardens, and admised of their beautiful flowers, fruits, leaves and stems. The plants suitable for avenues, pot-plants, and vase-flowers are also mentioned here.

596. **Magnolia obovata,** *Th.*, Jap. *Mokurenge, Shimokuren;* a garden tree of the order Magnoliaceæ attaining to a height of about 10 fts. It opens the flowers, dark purple on the outside and purple inside, before it sprouts. There is a variety called *Sarasa-renge* (M. obovata purpurea), with small light purple flowers.

597. **Magnolia conspicua,** *Salisb.*, Jap. *Haku-mokuren, Giokuran;* a garden plant of the order Magnoliaceæ, attaining to a height of 10-20 fts. It is nearly the same as the preceding in form, but the flowers are pale white, being prized for their fragrant odour.

597. b. **Magnolia stellata,** *Max.*, Jap. *Shide-kobushi, Hime-kobushi;* a species allied to Magnolia kobus (384). The flower consists of about 10 petals, which are narrow and pink-shaded white. There are still other varieties of Magnolia kobus, as called *Ōkobushi* (large) and *Murasaki-kobushi* (purplish).

598. **Magnolia parviflora,** *S. et Z.*, Jap. *Ōyama-renge;* a garden plant of the order Magnoliaceæ growing to a height of about 10 fts. It resembles Magnolia obovata (596) in shape, but smaller. The flower is white with red stamens, and is fragrant.

599. **Sterculia platanifolia,** *L.*, Jap. *Ao-giri, Itsu-saki;* a garden plant of the order Sterculiaceæ growing to a height of 40-50 fts. It produces male and female flowers, and yields pods which burst when ripe and expose small round edible seeds attached to both edges of the pods. Fibre is prepared from the bark. The mucilaginous substance contained in the bark is used

as a cement in pepar making. On account of its broad leaves and green stems, it is planted in gardens and as avenues.

600. Stuartia pseudo-camellia, *Max.*, Jap. *Natsu-tsubaki, Shara, Yama-kwarin;* a mountain tree of the order Ternstræmiaceæ attaining to a height of 20–30 fts. In summer it produces white single-petaled camellia-like flowers, whence the Japanese name. It is often planted in gardens, and the flowers are admired to be kept in a vase.

601. Acer japonicum, *Th.*, Jap. *Meigetsu-kayede, Hō-chiwa-momiji, Itaya-meigetsu;* a mountain-tree of the order Aceraceæ, attaining to a height of about 10 fts. In spring it sprouts at the same time with its male and female flowers, and yields seeds with samara. The leaves are large and furnished with many segments and are prized as one of the most beautiful maples on account of their red tint in autumn.

602. Acer trifidum, *Th.*, Jap. *Tō-kayede;* a garden plant of the order Aceraceæ, attaining to a height of 20–30 fts. The shape of the flowers and fruits are the same as the preceding. The leaves are ternate, and are prized for their yellow tint in autumn.

603. Acer polymorphum, *S.* et *Z.*, Jap. *Ichigiōji;* a mountain-tree of the order Aceraceæ attaining to a height of about 10 fts. The flowers and fruits are nearly the same as those of the preceding, but the fruits fall when fully ripe, and then the foliage becomes dark red. A kind with large leaves is called *Ō-sakadsuki.* There are several varieties of diverse forms and colours of leaves. The varieties originated from Acer polymorphum, A. japonicum, A. pictum, etc. are very numerous.

604. Acer polymorphum, *S.* et *Z.*, var., Jap. *Arisu-gawa-momiji, Beni-shidare;* a garden plant of the order Aceraceæ attaining to a height of 7–8 fts. Its branches have a drooping nature. This is distinguished for the unchanging dark red colour of its foliage with many dissected edges.

605. Hibiscus mutabilis, *L.*, Jap. *Fuyō, Kihachisu;* a garden shrub of the order Malvaceæ attaining to a height of about 10 fts. In cold places its stems die every winter, but it brings forth new stems in the next spring, growing to a height of 3-4 fts. In late summer it bears single or double and white or red beautiful flowers. Fibre is obtained from the bark, and also this plant is used in the same way as Hibiscus syriacus (329).

605. b. Hibiscus syriacus, *L.*, Jap. *Mukuge, Hachisu;* the fine flowers of this plant (329) are pink, white, or blue colour, and single or double.

606. Vitis inconstans, *L.*, Jap. *Nishiki-dsuta, Natsu-dsuta;* a climbing wild plant of the order Vitaceæ. In summer it shoots forth small peduncles from the axils of leaves, and bears many tiny flowers which are succeeded with black bean-sized round berries. Late in autumn, the foliage turns beautifully red, whence the Japanese name.

607. Evonymus alatus, *Th.*, Jap. *Nishikigi, Mayumi;* a mountain-shrub of the order Celastraceæ, growing to a height of 6-7 fts. Its stems and branches are furnished with longitudinal alate expansions. In summer it opens tiny flowers being succeeded with fruits which expose reddish yellow seeds when ripe. The foliage turns beautifully red in late autumn.

608. Evonymus oxyphyllus, *Miq.*, Jap. *Tsuri-bana;* a mountain-shrub of the order Celastraceæ, attaining to a height of 7-8 fts. The leaves and flowers resemble somewhat those of the preceding, but the fruits hang down at the ends of the long peduncles, exposing red seeds when ripe.

609. Evonymus tanakeii, *Max.*, Jap. *Koku-tengi;* a garden tree of the order Celastraceæ, attaining to a height of about 10 fts. When planted in warm regions its leaves do not fall off. Late in autumn the leaves turn reddish purple. The flowers and fruits are nearly the same as the preceding.

610. Milletia japonica, *A. Gray*, Jap. *Natsu-fuji, Doyō-fuji, Ko-fuji;* a climbing plant of the order Leguminosæ found wild in warm regions. The flowers, fruits, and leaves resemble those of Wistaria chinensis (334), but smaller. In mid-summer it opens white pale yellow flowers.

611. Wistaria chinensis, *S.* et *Z.*, var., Jap. *Noda-fuji;* a climbing plant of the order Leguminosæ with long panicles of flowers, 2–5 fts. long. It is produced in *Noda* in Province *Settsu,* whence the Japanese name.

611. b. Wistaria chinensis, *S.* et *Z.*, Jap. *Fuji, Yama-fuji;* the flowers of this climbing plant (334) are generally purple and single, but there is a variety with white and double flowers. They are used for a garden-ornament by letting creep over trellis.

612. Lespedeza buergeri, *Miq.*, var. intermedia, Jap. *Hagi;* a leguminous wild or planted shrub, growing in bushes from one root and attaining a height of 4–5 fts. Its flowers open in autumn, and their colours are white, purple, or reddish purple. A variety blooming in summer is called *Natsu-hagi.* There is another variety called *Miyagino-hagi,* which is very pretty with red flowers.

613. Cercis chinensis, *Bunge*, Jap. *Hanasuō, Suō-bana;* a garden plant of the order Leguminosæ attaining to a height of about 10 fts. In spring it opens reddish purple flowers in clusters before the leaves shoot forth, being succeeded with small pods.

614. Albizzia julibrissin, *Boivin*, Jap. *Nemu-no-ki, Kōka-no-ki;* a mountain tree of the order Leguminosæ growing to a height of about 10 fts., with bipinnate leaves. In summer it produces very ornamental flowers at the tops of the branches in the shape of a red tuft.

615. Prunus mume, *S.* et *Z.*, var., Jap. *Bun-yei-bai;* a garden shrub of the order Rosaceæ attaining to a height of 1–3 fts. It bears flowers and fruits in the next spring after sowing. The flowers are white and 5-petaled, and the fruits are larger than those of *Ko-mume* (167).

616. Prunus mume, *S.* et *Z.*, var., Jap. *Kōbai ;* its flowers are single or double and pink or dark red. The variety mentioned here is the common one. Besides this there are several other kinds of different colours and forms.

617. Prunus pseudo-cerasus, *Lindl.* fl. pleno., Jap. *Yaye-zakura ;* a garden tree of the order Rosaceæ attaining to a height of 10-20 fts. In spring before sprouting it bears many double light pink flowers, which are very fine. Besides the common one here mentioned, there are many varieties of white, red, yellow, or green and single or double flowers.

617. b. Prunus pseudo-cerasus, *Lindl.* fl. simple, Jap. *Hitoye-zakura ;* a variety of the preceding with pink or red single petaled flowers.

618. Prunus pseudo-cerasus, *Lindl.* fl. pleno., Jap. *Fugenzō ;* a variety of the preceding. In spring after sprouting double pink flowers appear. These flowers produce crimson new leaves at the centre.

619. Prunus subhirtella, *Miq.*, Jap. *Higan-zakura ;* a garden tree of the order Rosaceæ, growing several fts. high. It blooms before sprouting in spring, and the flowers are single and pink, being followed with small red berries, which turn dark purple when ripe.

620. Prunus subhirtella, *Miq.*, var. pendula, Jap. *Shidare-zakura ;* it has drooping branches, and the flowers and leaves are almost the same as the preceding, but more beautiful.

621. Prunus japonica, *Th.*, Jap. *Niwa-mume ;* a garden shrub of the order Rosaceæ attaining to a height of 20-30 fts. In spring it opens single small white flowers shaded with pink, being succeeded with small purplish red berries of a bitter and aciduous taste.

622. Prunus japonica, *Th.*, var. B. glandulosa, Jap. *Niwa-zakura;* its white or pink double flowers are very beautiful.

623. Prunus persica, *Benth* et *Hook*, Jap. *Momo;* this plant (184) is admired of its flowers in spite of its fruits. The common variety is mentioned here. There are several others with single or double and white or pink flowers.

624. Prunus persica, *Benth* et *Hook*, fl. rubra, Jap. *Hitō;* a variety of peach prized for its double, deep crimson, and long durable flowers.

625. Spiræa thunbergii, *Sieb.*, Jap. *Yuki-yanagi*, *Kogome-bana;* a mountain shrub of the order Rosaceæ growing in bushes 4–5 fts. high. In spring it bears small 5-petaled flowers in clusters before the leaves. They look like snow flakes, whence the Japanese name.

626. Spiræa prunifolia, *S.* et *Z.*, Jap. *Shijimi-bana*, *Haje-bana;* a garden shrub of the order Rosaceæ very much like the preceding in shape and quality, with round leaves and double white flowers in the form of balls.

627. Spiræa cantonensis, *Lour.*, Jap. *Kodemari;* a garden-shrub of the order Rosaceæ attaining to a height of 3–4 fts. In spring it bears small white flowers in umbels in the form of small balls.

628. Spiræa japonica, *L.*, Jap. *Shimotsuke;* a mountain-shrub of the order Rosaceæ attaining to a height of 3–4 fts. In summer it bears small flowers forming umbels. Their colours are white, pink, red, etc.

629. Kerria japonica, *D.C.*, Jap. *Yamabuki;* a wild shrub of the order Rosaceæ attaining to a height of 3–4 fts. In spring after sprouting it bears single or double yellow flowers. The variety of single flowers has several seeds on a calyx.

630. Rhodotypos kerrioides, *S.* et *Z.*, Jap. *Shiro-yamabuki;* a garden shrub of the order Rosaceæ attaining to a height of 4-5 fts. In early summer it bears 4-petaled white flowers on the branches, being succeeded with small round black seeds.

631. Rosa acicularis, *Lindl.* Jap. *Sakura-bara;* a garden shrub of the order Rosaceæ attaining to a height of 4-5 fts. In early summer it bears 5-petaled pink flowers, resembling those of the cherry, whence the Japanese name is derived. The variety blooming in all seasons is called *Kōshin-bara.* Besides this, there are *Kibara, Ukyōbara,* etc., with single or double and pink, white or yellow flowers. There are still numerous varieties lately introduced.

631. b. Rosa, Jap. *Goya-bara, Kaidō-bara, Ibara-shōbi;* a garden shrub of the order Rosaceæ resembling the wild rose (387) in shape, with its climbing stem. In summer it bears many red flowers forming a panicle. It is planted for hedges.

631. c. Rosa microphylla, *Roxb.,* Jap. *Sanshō-ibara, Izayoi-ibara;* a mountain-shrub of the order Rosaceæ attaining to a height of 5-6 fts. It is furnished with many thorns, and its leaves resemble those of Zanthoxyllum piperitum, whence the Japanese name is derived. The flowers are pink and double with lac on one side, whence it is also called *Izayoi-ibara.* Its fruits ripen in autumn and have an aciduous taste.

631. d. Rosa hystrix, *Lindl.,* Jap. *Naniwa-ibara;* a climbing rose with a long thorny vine. In summer it bears single white flowers about 3 inches in diameter, resembling somewhat those of Camellia; so it is also called Summer-camellia. There is a variety with pink and double flowers.

631. e. Rosa rugosa, *Th.,* Jap. *Hama-nasu;* a garden shrub of the order Rosaceæ attaining to a height of 2-3 fts. It grows wild on sandy ground near sea-shores in northern provinces. The flowers are generally single and red, but there are planted those with double and white flowers. They are very fragrant. The bark of the root is used for dying brown. In the district of *Akita* in Province *Ugo* it is used to dye *Hachijō*-silk.

632. Pyrus japonica, *Th.*, var. genuina, *Max.*, Jap. *Boke, Karaboke ;* a garden shrub of the order Rosaceæ attaining to a height of 6–7 fts. In spring before sprouting it bears pretty flowers, which are red, white, or variegated. The fruits are oval and 2–3 inches long, resembling Pyrus cydonia (191), but smaller It is used in the same way, and also as a medicine.

633. Pyrus japonica, *Th.*, var. pygmæa, *Max.*, Jap. *Kusa-boke, No-boke, Shidomi ;* a wild shrub of the order Rosaceæ attaining to a height of 1–2 fts. In early summer it bears red flowers. There is a garden variety with white flowers. Late in autumn, the fruits ripen and are edible.

634. Pyrus spectabilis, *Sit.*, Jap. *Kaidō ;* a garden-tree of the order Rosaceæ attaining to a height of about 10 fts. In spring it produces flowers on long peduncles. When they are still in buds, their colour is red, but when they open the outside of the petals is white and pink, and the inside is red. Sometimes they yield small round fruits a little larger than those of Nandina domestica.

634. b. Pyrus, Jap. *Nagasaki-ringo, Ko-ringo, Minari-kaido ;* a tree resembling very much the preceding, growing a little larger. It is used to graft the preceding. The flowers are larger and lighter. It yields small apple-like fruits, which are yellow when ripe and are edible.

635. Pyrus, Jap. *Rinki, Rinkin, Beni-ringo ;* a garden tree of the order Rosaceæ growing in cold regions, attaining to a height of 40–50 fts. Its flowers resemble apple-flowers, but smaller. The buds are red, and white when open. It yields many red fruits, which are good to eat.

636. Amelanchier asiatica, *C. Kock.*, Jap. *Shide-zakura, Zaifuri-boku ;* a mountain-tree of the order Rosaceæ found in warm regions growing to a height of about 10 fts. Late in spring, it produces hairy leaves and narrow white 5-petaled flowers disposed in short panicles, looking very pretty when waving by wind. The fruits are small and red when ripe.

637. Lagerstrœmia indica, *L.*, Jap. *Saru-suberi, Hiyakujikkō;* a garden-tree of the order Lythraceæ attaining to a height of about 10 fts. In late summer, it opens pretty red flowers in panicles. Other varieties with purplish or white flowers have been lately introduced. The trunk of this plant has a smooth bark, and it is said that even monkeys cannot climb up it, whence the Japanese name.

638. Deutzia scabra, *Th.*, Jap. *Utsugi, Uno-hana;* a wild shrub of the order Philadelphaceæ attaining to a height of 5-6 fts. In early summer, it produces 5-petaled white flowers. There is a variety with double pink flowers.

638. b. Deutzia sieboldiana, *Max.*, Jap. *Maruba-utsugi;* a pretty shrub resembling very much the preceding, with round leaves and early blooming white flowers.

638. c. Deutzia gracilis, *S. et Z.*, Jap. *Hime-utsugi, Chōsen-utsugi;* a smaller type of Deutzia scabra growing to a height of about 1 ft., with pretty closely clustered flowers.

638. d. Philadelphus coronarius, *L.*, var. satsumi, *Max.*, Jap. *Baikwa-utsugi, Fusuma-utsugi, Yoyogawa-utsugi;* a wild shrub with broad leaves and large white 4-petaled fragrant flowers.

639. Punica nana, *L.*, Jap. *Chōsen-zakuro, Nankin-zakuro;* a garden shrub of the order Myrtaceæ attaining to a height of 1-2 fts. When cultivated in a fertile soil it grows to a height of about 10 fts. It resembles very much P. granatum, though smaller, with single or double flowers which are deeper red.

639. b. Punica granatum, *L.*, var., Jap. *Hana-zakuro;*

it has double flowers, but no fruits. There exists another variety with white tips of petals. Both are pretty summer plants.

640. Hydrangia horteusis, *Smith*, var., Jap. *Ajisai;* a garden half lignous shrub of the order Saxifragaceœ growing in the form of a bush, attaining to a height of 4–5 fts. In early summer it bears flowers forming a large ball consisting of many pseudo-flowers with small real flowers hidden under them. They are white in the beginning, but turn blue and finally red. There are several sorts of the same nature.

640. b. Viburnum plicatum, *Th.*, Jap. *Temari-bana;* a variety of the preceding with white flowers forming a ball.

641. Corylopsis spicata, *S. et Z.*, Jap. *Tosa-midsuki;* a garden shrub of the order Hamameliaceœ attaining to a height of 7–8 fts. In spring it produces yellow drooping flowers with the calyx and peduncle of the same colour, being succeeded with bean-sized fruits.

642. Corylopsis pauciflora, *S. et Z.*, Jap. *Iyo-midsuki, Inu-midsuki, Kodosa;* a species of the preceding. The leaves and flowers are much alike, but smaller.

643. Hamamelis japonica, *S. et Z.*, Jap. *Mansaku;* a mountain-tree of the order Hamameliaceœ attaining to a height of about 10 fts. In spring before sprouting it bears flowers with short peduncles. The petals are very slender and of a golden colour, looking like golden threads. A variety called *Mume-dsuye* has smaller leaves and yellow flowers.

644. Liquidambar formosa, var. maximowiczii, Jap. *Fū;* a garden tree of the order Hamameliaceœ introduced in the year 1811. It attaints to a height of 20–30 fts. In spring it produces male and female flowers at the same time with the leaves. The female flowers are succeeded with prickly ball. The leaves turn red in late autumn, and are very pretty.

645. Cornus kousa, *Buerg.*, Jap. *Yama-bōshi, Karakwa,*

Itsuki ; a mountain-tree of the order Cornaceæ attaining to a height of about 10 fts. In summer it bears flowers in clusters in the form of a small ball, at which centre are four large white sepals looking like petals. The ball produces a red pulp which is edible and delicious.

646. Lonicera xylosteum, *L.*, Jap. *Hyōtan-no-ki, Kingin-boku ;* a mountain-shrub of the order Loniceraceæ attaining to a height of 5-6 fts. It blooms in early summer. The flowers are white at first, but turn yellow afterwards. The fruits have the form of gourds. They are red when ripe, but not edible.

647. Diervilla grandiflora, *S.* et *Z.*, Jap. *Hakone-utsugi, Nana-boke ;* a mountain-shrub of the order Caprifoliaceæ found much on the mountains of *Hakone,* whence the name. In summer it produces many flowers in clustered panicles on the branches. They are white in the beginning, but turn pink and then red.

648. Ehretia macrophylla, *Wall.*, Jap. *Mitsuna-gashiwa, Maruba-chisha, Tosa-giri ;* a mountain-tree of the order Boraginaceæ found in warm regions, attaining to a height of 20-30 fts. In summer it opens small yellowish white flowers in an umbel, being succeeded with bean-sized round fruits which are black when ripe. The leaves are thick and rough, being used for polishing purposes.

649. Styrax obassia, *S.* et *Z.*, Jap. *Hakuun-boku, Ōba-jisha ;* a mountain-tree of the order Styracaceæ attaining to a height of 20-30 fts. In summer it bears white flowers in panicles, being succeeded with drooping fruits, which give an oil.

650. Enkyanthus japonicus, *Hook*, Jap. *Dōdan, Dōdan-tsutsuji ;* a mountain-shrub of the order Ericaceæ attaining to a height of 7-8 fts. In spring it produces white bell-shaped small flowers drooping with long peduncles, and in late autumn its leaves turn red and are very beautiful.

651. Andromeda campanulata, *Miq.*, Jap. *Yōraku-dōdan, Yashio-tsutsuji;* a mountain-shrub of the order Ericaceæ attaining to a height of about 10 fts. It resembles the preceding in form, but the leaves are larger. Its pretty small crimson campanulate flowers droop down from the branches.

652. Rhododendron sinense, *Sweet,* Jap. *Ki-tsutsuji, Renge-tsutsuji, Ki-renge;* a wild shrub of the order Ericaceæ attaining to a height of 5-6 fts. In early summer it bears yellow flowers in clusters. A variety with yellowish red flowers is called *Kaba-renge.*

652. b. Rhododendron indicum, *Sweet,* var. kœmpferi, *Max.,* Jap. *Yama-tsutsuji;* a mountain-shrub of the order Ericaceæ, attaining to a height of 3-4 fts. In early summer it bears flowers on the branches resembling the preceding, but smaller. There are red and purple varieties, which latter is larger in general form and flowers, and blooms earlier.

652. c. Rhododendron rhombicum, *Miq.*, Jap. *Mitsuba-tsutsuji, Iwayama-tsutsuji;* a mountain-shrub of the order Ericaceæ attaining to a height of about 10 fts. Late in spring it bears purplish flowers, and when the flowers decay it produces three leaves in a circle, whence the name (three-leaved Azalea) is derived.

652. d. Rhododendron ledifolium, *Don.,* Jap. *Neba-tsutsuji, Mochi-tsutsuji;* a mountain-shrub of the order Ericaceæ resembling *Yama-tsutsuji* (652. b.), attaining to a height of 3-4 fts. In early summer it bears purplish flowers on the branches, provided with a sticky substance on the peduncles, whence the Japanese name (sticky Azalea) is derived. A variety with narrow leaves and petals is called *Seigai-tsutsuji.*

653. Ilex sieboldii, *Miq.*, Jap. *Mume-modoki;* a mountain-shrub of the order Aquifoliaceæ growing to a height of about 10 fts. In summer it opens small flowers, being succeeded with small round berries which turn red or white when ripe in winter

654. Jasminum sieboldianum, *Blume.*, Jap. *Ōbai;* a garden shrub of the order Jasminaceæ with a slender vine-like stem, being several feet long. In early spring it bears pure yellow flowers before the leaves, and is one of the trees which bloom very early. There is a variety called summer Jasmin with persistant leaves.

655. Tecoma grandiflora, *Delaun.*, Jap. *Nōzen-kadsura;* a garden climber of the order Bignoniaceæ. In late summer it produces panicles with several orange red flowers.

656. Callicarpa japonica, *Th.*, Jap. *Yabu-murasaki, Mi-murasaki;* a wild shrub of the order Verbenaceæ attaining to a height of several feet. In summer it produces small purple flowers in clusters, being succeeded with small round purple berries. There are different sorts called *Ko-murasaki* (small purple), *Yama-murasaki* (mountain purple), etc.

657. Salix babylonica, *L.*, Jap. *Shidare-yanagi;* a garden tree of the order Amentaceæ attaining to a height of 3-4 fts. The branches are slender and drooping to the ground. It is a diœcious plant, blooming in spring before the leaves. The female flowers when ripe disperse white cotton-like fibres. The variety here mentioned is *Rokkakudo* which shoots forth long drooping branches. This plant thrives well in moist places, and is suitable for avenues and gardens. It is also used for vase-flowers.

658. Salix buergeriana, *Miq.*, Jap. *Neko-yanagi, Kawa-yanagi, Saru-yanagi;* a wild tree of the order Amentaceæ attaining to a height of about 10 fts. It resembles the preceding in aspect, but the branches do not droop. The female flowers are covered with soft silky hair, whence the name *Neko-yanagi* (cat willow) is derived. The branches are used for vase-flowers. The ripe seeds produce cotton-like fibre.

658. b. Tamarix chinensis, *Lour.*, Jap. *Gyoryū;* a garden tree of the order Tamaricaceæ attaining to a height of about 10 fts. The leaves look like needles, and the branches droop like the weeping willow. In summer and autumn it produces panicles of small pink flowers.

658. c. Forsythia suspensa, *Vahl.*, Jap. *Rengiō, Itachi-kusa, Itachi-base;* a garden shrub of the order Oleaceæ with slender drooping branches. In spring it bears 4-lipped tubular yellowish flowers before sprouting, being succeeded with heart-shaped fruits.

658. d. Stachyurus præcox, *S.* et *Z.*, Jap. *Ki-fuji, Mame-fuji;* a mountain-shrub of the order Ternstrœmiaceæ attaining to a height of 8–9 fts. In spring it produces panicles of small yellow flowers 3–4 inches long, being succeeded with bean-sized fruits.

658. e. Citrus trifoliata, *L.*, Jap. *Karatachi, Kikoku;* a garden shrub of the order Aurantiaceæ attaining to a height of 5–10 fts. In late spring it bears 5-petaled white flowers, being succeeded with small round oranges, which are yellow when ripe. They are not eatable on account of their bitterness and acidity. The plants are used as stocks for grafting orange trees, and also for hedges.

Note.—Many other deciduous garden plants are contained in the chapters of fruits, oil and wax, textile and paper-manufacturing plants, dying, fragrant, medicine, and timber trees, etc., but they are omitted here.

VOLUME III.

CHAPTER XXIII.—EVERGREEN GARDEN-TREES AND SHRUBS.

This Chapter includes all ornamental plants which leaves do not fall in any season. These are planted in gardens as ornaments on account of their green leaves and beautiful yellow or white variegation, and also for the beauty of their flowers. Besides these, some used for hedges, pot-plants, or vase-flowers are also mentioned here.

659. Magnalia compressa, *Max.*, Jap. *Ogatama-no-ki;* a tree growing in mountains of warm regions, attaining to a height of about 10 fts. In early summer it bears white flowers slightly shaded with pink, yielding fruits which expose red seeds when fully ripe.

660. Berberis japonica, *R. Br.*, Jap. *Hiragi-nanten* a garden shrub of the order Berberidaceæ growing to a height of 3-4 fts. In early summer, it produces many small yellow flowers attached to a long stalk shooting out of the axils of leaves. After blooming, small dark purple berries are produced. Its leaves resemble those of Nandina domestica in form, and as its dented edges resemble somewhat those of Olea aquifolium, whence the Japanese name.

661. Ternstrœmia japonica, *Th.*, Jap. *Mokkoku;* a garden tree of the order Ternstrœmiaceæ growing wild on seashores of warm regions. It attains to a height of about 20 fts. In the beginning of summer white flowers come forth, being followed with small fruits which burst and expose red seeds when ripe. Its reddish brown wood, being finely grained, is used to make several articles, and its bark is used for tincture.

662. Cleyra japonica, *Th.*, Jap. *Sakaki, Masakaki;* a mountain-tree of the order Ternstrœmiaceæ attaining to a height of about 10 fts. Early in summer, yellowish white small flowers shoot forth, being followed with round berries. This tree is generally offered before gods.

663. Eurya japonica, *Th.*, Jap. *Hisakaki, Mesakaki, Shirashake;* a mountain-tree of the order Ternstrœmiaceæ attaining to a height of about 10 fts. Late in spring, small greenish white flowers appear on the axils of leaves, being succeeded with small berries which turn dark purple when ripe.

664. Camellia japonica, *L.*, var. aquifolium, Jap. *Hiiragi-tsubaki;* its leaves are dented, resembling those of Olea aquifolium. It is planted in gardens, growing to a height of 6–7 fts. Its flowers are light pink or white.

665. Camellia japonica, *L.*, var , Jap. *Otome-tsubaki;* it grows in gardens to a height of 6–7 fts. Its small plants suit to plant in pots. Its flowers are double and pink or white, being esteemed for their beauty. Besides this, there are many varieties of Camellia, but they are omitted here.

666. Camellia sasanqua, *Th.*, Jap. *Sazankwa, Ko-tsubaki;* a garden or mountain tree of the order Ternstrœmiaceæ, resembling very much Camellia japonica in form, though smaller, growing to a height about 10 fts. The flowers appear late in autumn, and they are single petaled and of several colours, pink, white, variegated, etc. Oil is taken from the seeds as in Camellia japonica.

667. Pittosporum tobira, *Ait.*, Jap. *Tobera;* a mountain-tree of the order Pittosporaceæ growing to a height of about 10 fts. Early in summer, it produces yellowish white flowers forming a raceme among the leaves. The flowers are succeeded with round berries which expose red seeds when ripe.

668. Ilex crenata, *Th.*, Jap. *Inu-tsuge;* a mountain tree of the order Aquifoliaceæ growing to a height of about 10 fts. In early summer, small yellowish green flowers open, being succeeded with small round dark purple berries.

669. Ilex latifolia, *Th.*, Jap. *Tarayō;* a mountain tree of the order Aquifoliaceæ, growing to a height of 20–30 fts. In summer small greenish flowers come forth in bunches, being followed with small red berries.

669. b. Ilex integra, *Th.*, Jap. *Mochi-no-ki;* this tree (303) is much planted in gardens or for hedges.

670. Evonymus japonicus, *Th.*, Jap. *Masaki;* a wild tree of the order Celastraceæ growing to a height of about 10 feet, being used generally for hedges. Various kinds of variegated leaves exist. In early summer small flowers open on the branches, and in late autumn the berries expose red seeds.

671. Raphiolepis japonica, *S. et Z.*, Jap. *Hamamokkoku;* a wild shrub of the order Rosaceæ found on sea shores of southern provinces. It attains to a height of 3–4 fts., branching out horizontally. In early summer, it bears white flowers which resemble those of Prunus mume, being succeeded with dark purple berries. Its bark is used for dying (374. b.).

672. Photima glabra, *Th.*, Jap. *Kaname-mochi, Akame-mochi, Kaname-gashi, Soba-no-ki;* a mountain tree of the order Rosaceæ growing to a height of about 10 fts. Its small trees suit well for hedges. In early summer small white flowers in a terminal cyme appear on the branches, being succeeded with small round red berries. As the young leaves are red, they are called *Aka-me* (Red-shoots). The wood, being very hard and finely grained, is used for wheels and oars in Province *Kii.*

673. Fatsia japonica, *Decne. et Planc.*, Jap. *Yatsude, Tengu-no-hauchiwa;* a mountain shrub of the order Araliaceæ growing in warm provinces. It attains to a height of 7–8 fts., shooting stems in tufts from a root. In winter, branched peduncles are produced in the centre of the leaves, bearing small yellowish white flowers forming round balls, being followed with black berries.

674. Dendropanax japonicum, *Seem.*, Jap. *Kakure-mino, Mitsude, Kara-mitsude, Miso-buta ;* a mountain tree of the order Araliaceæ growing in warm provinces, attaining to a height of about 20 fts. Late in autumn, it yields flowers. It resembles the preceding in all respects, though its berries are a little smaller.

675. Hedera helix, *L.*, Jap. *Fuyu-dsuta, Ki-dsuta ;* a plant of the order Araliaceæ climbing on other trees or stones, or creeping on the ground. In some large vines, the stems are about 3-4 inches thick. During winter it blooms being succeeded with fruits, which resemble those of Fatsia japonica, though smaller in size. A variety with dentate leaves is called *Momiji-dsuta*.

676. Aucuba japonica, *Th.*, Jap. *Aokiba, Aoki ;* a mountain shrub of the order Cornaceæ growing to a height of 7-8 fts. It is a diœcious plant. In late spring it shoots young stalks on the branches, bearing small purplish green flowers. Its fruits are red when ripe in winter. The leaves are white or yellow variegated or margined, and some are narrow. It thrives well in shady places.

677. Rhododendron brachycarpum, *Don.*, Jap. *Shakunange, Shakunagi ;* a mountain shrub of the order Ericaceæ attaining to a height of 7-8 fts. Those growing on lofty mountains creep on the ground. Their large and thick leaves grow closely at the tops of the branches, and in their centre beautiful light pink flowers appear in clusters, looking like peony-flowers at distance.

677. b. Rhododendron indicum, *Sweet*, var. obtusum, *Max.*, Jap. *Kirishima-tsutsuji ;* a mountain shrub of the order Ericaceæ attaining to a height of 3-4 fts. and sometimes about 10 fts. The leaves are small and the red flowers, which bloom all at the same time, are pretty. There are different varieties of flowers, small or large, single or double, white or purple, etc.

677. c. Rhododendron indicum, *Sweet*, var. macranthum, *Max.*, Jap. *Satsuki-tsutsuji;* it grows to a height of 3-4 fts. Its flowers are a little larger than the preceding. The flowers are generally red or purple, but sometimes white. The flowers open in May.

677. d. Rhododendron sablanceolatum, *Miq.*, Jap. *Riukiu-tsutsuji;* it resembles the preceding in form, but larger. The flowers are white, and sometimes light purple.

678. Ligustrum ciliatum, *Sieb.*, Jap. *Iwaki;* a garden shrub of the order Oleaceæ attaining to a height of 5-6 fts In early summer, it bears many small white flowers disposed in panicles, being succeeded with round dark purple berries.

679. Daphniphyllum macropodum, *Miq.*, Jap. *Yudsuriha ;* a mountain tree of the order Euphorbiaceæ attaining to a height of about 10 fts. It is a diæcious plant, and in early summer small yellowish green flowers appear stalks produced at the centre of the leaves, being followed with small oval black fruits. The leaves are used for a congraturating decoration on the new year's days.

680. Ficus pyrifolia, *Poir.*, Jap. *Inu-biwa, Ko-ichijiku;* a wild tree of the order Urticaceæ, growing on sea-shores of warm regions, attaining to a height of about 10 fts. In late summer, it gives round fruits about the size of a thumb. When fully ripe, the fruits are dark purple, and edible. The bark is used for paper manufacture. The variety drawn in this book is the evergreen kind of Bonin Island.

681. Quercus sessilifolia, *Bl.*, Jap. *Tsukubane-gashi;* a mountain tree of the order Amentaceæ attaining to a height of about 30 fts. It resembles Quercus acuta (564), but its leaves shoot forth straightly in clusters.

682. Quercus phyllireoides, *A. Gray*, var., Jap. *Chirimen-gashi, Biwayō-gashi;* a variety of *Ubame-gashi* (566), but its leaves are wrinkled.

683. Quercus thalassica, *Hance*, var., Jap. *Shimagashi*, *Mokume-gashi*; a variety of *Shira-kashi* (565), but its leaves are variegated. It is planted in gardens.

684. Quercus lacera, *Bl.*, Jap. *Hiryō-gashi*; a species of Japanese evergreen oak. Its leaves are provided with acute narrow dents on the edges.

685. Quercus pinnalifida, *Fr.* et *Sav.*, Jap. *Hagoromo-gashi*; a kind of Japanese evergreen oak, named by the form of its leaves.

686. Pinus densiflora, *S.* et *Z.*, var., Jap. *Shiraga-matsu*; its leaves have white variegation at their extremities. Besides this, a variety with white variegation in the middle part of leaves is called *Ja-no-me*, a variety with entirely white leaves *Shimo-furi-matsu*, and a variety with yellow leaves *Ōgon-matsu*. There are still several other varieties with different forms of leaves.

687. Pinus koraiensis, *S.* et *Z.*, Jap. *Chosen-goyō*, *Kanshō*, *Chosen-matsu*; a garden tree of the order Coniferæ growing to a height of about 30 fts. with 5 long needle-leaves. Its cones are large, being 7–9 inches, and their nuts are edible (217. c.).

688. Pinus parviflora, *S.* et *Z.*, Jap. *Goyō-matsu*, *Shimo-furi-goyō*; a garden tree of the order Coniferæ attaining to a height of 20–30 fts. There are also dwarf trees suited for artificial garden-mountains and pot-plants. Those growing in high mountains creep over the ground, with short leaves.

689. Cunninghamia sinensis, *R. Br.*, Jap. *Kōyōsan*, *Riuhi*, *Oranda-momi*; a garden tree of the order Coniferæ growing to a height of 30–40 fts. In late spring, male and female flowers open, and afterwards cones are produced at the extremities of the branches. The pointed leaves grow pinnately on the branches, and sting the hand when touched.

Juniperus chinensis *L. var.* **procumbens.**
690. Jap. *Haibiyakushin;*
a garden shrub of the order Coniferæ. Its stems creep over the ground for the length of several feet.

691. Juniperus chinensis, *L.,* *Biyakushin, Tachi-biyakushin, Sugi-biyakushin;* a conifer resembling the preceding, but standing straight to a height of about 10 fts.

692. Biota orientalis, *Endl.,* Jap. *Konote-gashiwa;* a garden coniferous shrub attaining to a height of 6-7 fts. It grows ni a conical form, and its branches and leaves are arranged regularily. The leaves have no distinction on both sides and stand straight.

693. Thuya obtusa, *B. et H.,* var., Jap. *Kamakura-hiba;* a beautiful garden-tree.

694. Thuya obtusa, *H. et B.,* var., Jap. *Chabo-hiba;* its leaves and branches are very short.

695. Thuya obtusa, *H. et B.,* var., Jap. *Kujaku-hiba;* its leaves resemble peacock's feathers, whence its name is derived. A variety *Ōgon-kujaku* (golden peacock) is a little larger in form.

696. Thuya pisifera, *S. et Z.,* var., Jap. *Shinobu-hiba;* its leaves are fine and beautiful.

697. Thuya obtusa, *S. et Z.,* var., *Yenbi-hiba;* it has long drooping branches, which almost reach to the ground.

698. Chamæcyparis lycopodioides, var., Jap. *Seiriū-hiba;* its branches are long and drooping.

699. Thuya pendula, *Max.,* Jap. *Hiyoku-hiba;* it has drooping branches and leaves. There is a variety with white variegated leaves.

700. Cryptomeria japonica, *Don.,* var., Jap. *Ōgon-sugi;* it is admired of its beautiful light yellow leaves.

701. Cryptomeria japonica, *Don.*, var., Jap. *Yenkō-sugi;* its long extending branches are like monkey-arms.

702. Cryptomeria japonica, *Don.*, var., Jap. *Yore-sugi;* its branches and leaves are twisted.

703. Cryptomeria japonica, *Don.*, var., Jap. *Gorō-sugi;* its leaves grow in a beautiful compact order.

704. Cryptomeria japonica, *Don.*, var., Jap. *Bandai-sugi;* its short leaves grow together and form dense balls on the branches.

705. Cryptomeria japonica, *Don.*, var., Jap. *Aya-sugi;* its leaves grow twisted on the branches.

706. Cryptomeria japonia, *Don.*, var., Jap. *Yawara-sugi;* its leaves are fine and soft.

706. b. Chamæcyparis squarrosa, *S.* et *Z.*, Jap. *Hi-muro, Hime-muro;* its leaves resemble those of the preceding, but finer and shorter. It grows to a height of about 10 fts. It is closely allied to *Sawara*.

707. Sciadopytis verticillata, *S.* et *Z.*, Jap. *Kōya-maki, Kusa-maki;* a coniferous mountain tree growing high. In summer it bears male and female flowers, and afterwards cones about the size of a boy-wrist. It is esteemed for its splendid appearance in every season, with its stately ramified branches and umbrella-like arranged leaves. The wood is endurable for moisture, and its peeled bark is used by the name of *Maki-hada*.

707. b. Podocarpus macrophylla, *Don.*, Jap. *Maki, Inu-maki;* it is planted in gardens, and also used for hedges.

708. Cephalotaxus drupacea, *S.* et *Z.*, var., Jap. *Chōsen-gaya;* its leaves resemble the preceding, but smaller.

709. Taxus tradiva, *Laws*, Jap. *Kyara-boku;* it resembles *Ichii* (580). The stems are standing or creeping.

710. Cycas revoluta, *Th.*, Jap. *Sotetsu;* it grows in warm regions, attaining to a height of about 10 fts. and branching in groups. Its stem is covered with scales, and is highly prized as an ornamental garden or pot-plant on account of its beautiful slender lanceolate leaves growing pinnately. The male inflorescence of this diæcious plant comes forth in the middle stem in the form of a club, being 2-3 fts. long, while the female flowers open on the hand-like stalks, to which the seeds are attached. The seeds are about the size of a small peach, and are of a bright colour. The kernels are edible when broiled (217. c.). Starch is obtained from its stem. Its leaves are used for plaiting hats and baskets.

711. Chamærops excelsa, *Th.*, Jap. *Shuro;* it is produced in warm regions. Its stem attains to a height of 30-40 fts., growing straight in the form of a club. The leaves grow at the summit of the stem, and expand like fans on long peduncles. Male and female flowers grow on separate plants. The male flowers are yellow and millet grain sized, and the fruits of the female flowers are bean-sized. The outside of the stem is covered with hair called *Shuro-no-ke*, which is used for many purposes (349. c.). Its stem is also used as an ornamental wood.

712. Rhapis flabelliformis, *Ait.*, Jap. *Shūro-chiku;* an ornamental palm produced in warm regions. Its stems grow in groups, attaining to a height of about 10 feet. In cold regions, they do not grow so high, but only about 1 ft. The leaves are fan-shaped and deeply cut into segments. Its flowers open in spikes spreading into branches. Like the preceding, it is a diæcious plant. Its fruits are small and round with scales. Its stems being tough and strong are made into sticks, umbrella-handles, etc.

713. Rhapis kwannontik, Jap. *Kwannon-chiku, Riu-kiu-shūro-chiku;* a palm resembling very much the preceding, but the stem is covered with much more fibres, and its dark green leaves are short and stiff.

714. — Phyllostachys heterocycla, *Carr.* _, Jap. *Kikkō-chiku;* a variety originated from B. mosa (590), with its knots attaching alternately by one another, forming tortoise-shell-figures of a length of 1-2 fts. upon the ground.

715. Bambusa sterilis, *Krz.*, Jap. *Hotei-chiku, Gosan-chiku;* an ornamental and useful plant of the order Gramineæ, growing abundantly in warm regions, attaining to a height of about 10 fts. The knots of the lower part of its stem are very narrow and irregular, forming a curious appearance. It is very suitable as a fishing rod, and also used as a stick and an umbrella-handle.

716. Bambusa marliacea, Jap. *Shibo-chiku;* it is principally produced in Province *Awaji*. It attains to a height of about 10 fts. with a diameter ½-2 inches. It resembles *Madake* (589) with longitudinal wrinkles on its stem, which give it a gracious appearance, being used to make flower vases and for other such purposes.

717. Bambusa pygmæa, *Miq.*, Jap. *Kimmei-chiku;* a garden bamboo growing to a height of about 10 fts. with a diameter of about 1 inches, resembling *Madake* (589) in shape. The stem is green on the side where the branches shoot forth, and the reverse side is yellow. The leaves have white stripes, and are very pretty.

718. Bambusa sp., Jap. *Han-chiku;* a mountain bamboo with many varieties. The variety drawn in this book is that much planted in the provinces of *Omi* and *Tamba*. The diameter of the stem is about 1½ inches. It has cloud-like variegation.

719. Bambusa quadrangularis, Jap. *Shikaku-dake, Shihō-chiku;* a garden bamboo growing to a height of about 10 fts., with a quadrangular stem, which diameter is about an inch. The leaves are small and narrow resembling those of *Madake*. It is one of the most strange varieties.

720. Bambusa sp., Jap. *Narihira-dake;* a garden bamboo attaining to a height of about 10 fts. Its stem resembles that of *Madake* (589), and the leaves resemble those of *Medake* (591). Its sprouts appear in May, and are provided with thick strong sheaths.

721. **Bambusa sp.**, Jap. *Tō-chiku ;* a garden bamboo having a stem attaining to a height of about 10 fts. with a diameter of about 1 inch. The stems and leaves resemble the preceding, but the leaves somewhat thinner, and the stems lighter coloured. The distance between every knot is very long. The branches grow in thick clusters.

722. **Bambusa sp.**, Jap. *Kan-chiku ;* a garden bamboo attaining to a height of 5-6 fts. It is used for hedges. Its full grown height is about 10 fts. with a diameter of about 1 inch. The stem is purplish in the upper part. From late autumn to winter, it produces the young sprouts, which are good to eat.

Phyllostachys bambusoides, *S. et Z.*
723. Jap. *Ya-dake, Ya-shino ;* a wild bamboo attaining to a height of 7-8 fts. The leaves are broad and large, resembling those of *Kuwa-zasa* (732). As the stem is slender and strong, it is used to make arrows. It is also useful for making baskets, sieves, etc.

Bambusa vulgaris, *Wendl.*
724. Jap. *Taisan-chiku, Tōgin-chiku ;* a garden bamboo attaining to a height of 20-30 fts., with a diameter of about 2½ inches. The stem is deep green with low even knots, and the leaves are broad, being very beautiful. It is easily propagated by cutting, but as it is originated in warm regions, it is often injured by cold.

Arundinaria Hindsii, *Munro.*
725. Jap. *Taimin-chiku, Daimyo-dake ;* a garden bamboo attaining to a height of about 10 fts., with a diameter of about 1 inch. The leaves and branches are slender and fine. The small plants are planted in pots or between rocks, and sometimes in water basins. The stems are suitable to make flutes.

726. **Bambusa sp.**, Jap. *Kanzan-chiku ;* a garden bamboo attaining to a height of about 10 fts., with the stem about an inch

in diameter, resembling *Medake* in shape. Whips are made of its roots. The characteristic nature of this bamboo is that the leaves are erect and the branches straight upwards. It is planted in pots or in water vases.

727. **Bambusa nana,** *Roxb.* Jap. *Hōō-chiku ;* a garden bamboo attaining to a height of 4-5 fts. It is suitable for hedges, and the small ones are planted in pots. The stems grow in tufts, and its leaves grow closely together resembling a bird's tail. There is a variety with golden yellow longitudinal stripes on its stem, and also a variety with yellow and white stripes on its leaves. All these are the varieties of *Usen-chiku* (349) and *Doyō-chiku* (728).

728. **Bambusa sp.,** Jap. *Doyō-chiku, Chin-chiku, Kin-chiku ;* the same species as *Usen-chiku* (349). As the stems grow in clusters, they are used for hedges. The character of this bamboo is that its underground stem creeps with close knots and the stem above the ground is straight with apart knots. It is used for a stick, umbrella-handle, etc. As the sheath is thick and very lustrous inside, it is used as a spoon.

729. **Bambusa sp.,** Jap. *Suō-chiku ;* a garden bamboo attaining to a height of 4-5 fts. Its young stem is red with green longitudinal stripes.

730. **Bambusa sp.,** Jap. *Shakotan-chiku ;* a mountain bamboo, being a kind of *Kuma-sasa*. The stem attains to a height of about 10 feet, and the part covered with sheash has a dark purple variegation. Its origin is in the districts of *Shakotan* in *Hokkaidō*, whence its name is derived. From its variegation it is also called *Shako-han-chiku* (partridge variegated bamboo). As the stems are tough and strong, they are used to make Japanese pen-holders, tobacco-pipes, sticks, and other articles.

731. **Bambusa sp.,** Jap. *Yakiba-zasa ;* a mountain bamboo, being a variety of *Kuma-zasa*, attaining to a height of 3-4 fts. with broard yellowish white margined leaves. It is very

ornamental in gardens, and its leaves are used for various purposes. **Bambusa veitchii,** *Carr.*

732. _ _ . Jap. *Kuma-zasa;* a famous mountain bamboo attaining to a height of 5-6 fts., with broad leaves about 8-9 inches long. A large variety called *Oni-kumazasa* is about 10 fts. high, and a small one called *Ko-kumazasa* is about 1 ft. high. The stems are slender and strong. It rarely gives grains, which are edible (20. c.).

Phyllostachys Kumasasa, *Munro.*

733. _ , Jap. *Bungo-zasa, Okame-zasa, Tōba-zasa, Iyo-zasa;* a mountain bamboo growing to a height of 3-4 fts. The stem is very slender with elevated knots, and the leaves shoot from each knot in a five-leaved-cluster. As its stems grow in groups, it is fitted for hedges. The stems are used to make baskets, etc.

Note.—Besides those mentioned here, there are many other plants which many be included in this chapter. The principal ones are as follows:—in the Chapter of Timber Trees and Bamboos, Distylium racemosum (544), Olea aquifolium (551), Cinnamomum camphora (553), Buxus japonica (556), Quercus acuta (564), Q. glauca (565), Q. phyllioides (566), Juniperus chinensis (572), Thuja dolabrata (573), T. japonica (574), Chamæcyparis obtusa (575), C. pisifera (576), Cryptomeria japonica (577), Podocarpus chinensis (578), P. nagæa (579), Taxus cuspidata (580), Pinus thunbergii (581), P. densiflora (582), P. parviflora (583), Abies firma (585), A. tsuga (586), Bambusa puberula (592), Phyllostachys nigra (590), etc.; in the Chapter of Fruit-trees Photinia japonica (192), Elœgnus longipes (213), Torreya nucifera (217), Quercus cuspidata (225), and Q. glabra (226); all kinds of oranges, which are good to plant in pots; in the Chapter of Plants for Luxury, Thea chinensis (283) and Ligustrum japonicum (290); in the Chapter of Oil and Wax Plants, Camellia japonica (311) and Cephalotaxus drupacea (314); in the Chapter of Dye-plants, Myrica rubra (374); in the Chapter of Odorous Plants, Olea fragrans (392) and O. fragrans (393);

and in the Chapter of Poisonous Plants, Illicium religiosum (484. b.), etc.

CHAPTER XXIV.—Ornamental Plants.

This chapter contains all the trees, shrubs, or herbs, which are planted in gardens or flower-beds as ornaments, and also suitable for pot-plants or vase-flowers. All their flowers and leaves are very beautiful and graceful.

734. Clematis patens, *Morr.*, et *Decne.*, Jap. *Tessen;* a climbing plant of the order Ranunculaceæ. In spring young shoots sprout from the old vines, and in summer blue flowers with small purple petals in the centre are produced, being about 2 inches in diameter. There is another variety which yields white flowers with narrow purple petals in the centre.

735. Clematis florida, *Th.*, Jap. *Kazaguruma;* a variety of the preceding, but it does not have purple petals in the centre. There are blue and white varieties. The variety with double white flowers is called *Yuki-okoshi*, and the variety with double blue flowers *Ruri-okoshi*.

736. Anemonopsis macrophylla, *S.* et *Z.*, Jap. *Kusarenge, Renge-shōma;* a perennial mountain herb of the order Ranunculaceæ growing to a height of about 2 fts. In summer each peduncle bears a white flower shaded with purple, resembling that of lotus.

737. Anemone japonica, *S.* et *Z.*, Jap. *Shumei-giku, Kibune-giku;* a perennial wild herb of the order Ranunculaceæ growing to a height of about 2 fts. In autumn it shoots forth petioles with flowers at the tops. The flower is reddish purple, resembling that of chrysanthemum.

738. Anemone cernua, *Th.*, Jap. *Okina-gusa, Shagmasaigo;* a perennial wild herb of the order Ranunculaceæ. In spring it bears purplish red flowers, being followed with many white fruits.

739. Anemone hepatica, *Gort.*, Jap. *Misumisō, Yukiwarisō, Suhama-saishin;* an evergreen mountain herb of the order Ranunculaceæ. It produces one flower at the top of each peduncle. The flowers are of different colours, as red, purple, and white. It is often planted in hot houses for flowers in spring.

740. Caltha palustris, *L.*, var. sibirica, *Reg.*, Jap. *Yenkōsō ;* a perennial herb of the order Ranunculaceæ growing wild in marshy places. In late spring, it shoots oblique peduncles with yellow flowers. There is a variety called *Riukinkwa* with its peduncles growing straight upwards.

741. Adonis amurensis, *Reg.* et *Radd.*, Jap. *Fukujisō ;* a perennial herb of the order Ranunculaceæ growing in northern provinces. In spring peduncles with bright yellow flowers are produced at the same time with the leaves. It is planted in hot houses for selling in early spring. There are many varieties.

742. Aquilegia glandulosa, *Fisch.*, Jap. *Odamaki, Odamakisō ;* a perennial garden herb of the order Ranunculaceæ. In late spring it bears single purplish blue or double white flowers at the tops of peduncles. The mountain variety is called *Yamaodamaki*.

743. Pœonia albiflora, *Pall.*, Jap. *Shaku-yaku, Kaoyogusa ;* a perennial garden herb of the order Ranunculaceæ. In spring it shoots forth stems, and in summer it bears flowers, which are single or double and red or white. The roots are used as medicine. There is also a mountain variety.

744. Pœonia mautan, *Sims.*, Jap. *Botan, Hatsuka-gusa ;* a deciduous garden shrub of the order Ranunculaceæ. In spring it sprouts and bears single or double flowers of diverse colours, red, pink, etc. The roots are used as medicine, and the flowers are edible.

745. Aceranthus diphyllus, *Morr.* et *Decne.*, Jap. *Baikwa-ikarisō;* a perennial wild herb of the order Berberideæ. In late spring, it shoots forth its stalks and yields flowers arranged in panicles. The flowers are pinkish white, and some-what resemble plum flowers in shape, whence the Japanese name.

746. Epimedium violaceum, *Morr.* et *Dec.*, Jap. *Ikarisō;* a perennial wild herb of the order Berberideæ. In spring it produces branched stalks bearing white or reddish purple anchor-shaped flowers.

747. Nandina domestica, *Th.*, Jap. *Nanten;* an evergreen shrub of the order Berberideæ growing wild in southern provinces. Generally the stem grows to a height of 4–5 fts., but it sometimes attains to 10 fts. It produces small white flowers disposed in panicles being followed with round red or white berries.

748. Nandina domestica, *Th.*, var., Jap. *Kinshi-nanten;* a dwarf variety of the preceding, growing to a height of 6–12 inches. The branches and leaves are very fine. There are many varieties.

749. Nelumbo nucifera, *Gærtn.*, Jap. *Hasu;* a perennial herb of the order Nymphæaceæ. Its roots and seeds are edible as described in 125 and 228. In summer it produces long peduncles above water-surface, bearing flowers on the tops. The flowers are of several sizes and colours, as white, pink, etc. As the flowers are very pretty, it is planted in ponds or basins.

750. Nelumbo nucifera, *Gærtn.*, var,, Jap. *Chawan-basu;* a dwarf variety of the preceding, blooming well in small basins.

751. Nymphæa tetragona, *Bemerl.*, Jap. *Hitsuji-kusa;* a perennial herb of the order Nymphæaceæ growing in marshes and ponds. Its leaves float on the surface of water. In middle summer it produces double white flowers which open afternoon.

752. Nuphar japonicum, *Dc.*, Jap. *Kō-hone ;* a perennial herb of the order Nymphæaceæ growing in marshes and ponds. Its leaves are above water-surface. In late summer it yields one yellow flower at the top of a peduncle. There are several varieties.

753. Papaver somniferum, *L.*, Jap. *Keshi ;* a biennial garden herb of the order Papaveraceæ, attaining to a height of 4-5 fts. In early summer it opens flowers of various colours. Opium is made from the young fruits. Its young leaves are edible when boiled.

754. Papaver rhœas, *L.*, Jap. *Hina-geshi, Bijin-sō ;* a dwarf variety of the preceding, with hairs on its stem and leaves. The stem attains to a height of 1-2 fts., and has only one flower on its top. There are several varieties as the preceding.

755. Stylophorum japonicum, *Miq.*, Jap. *Yamabuki-sō, Kusa-yamabuki ;* a perennial herb of the order Papaveraceæ growing wild in shady places. It blooms in late spring, and the flowers are yellow, resembling those of *Yamabuki*, whence its Japanese name. There is another variety with leaves, resembling those of *Seri*.

756. Dicentra spectabilis, *Miq.*, Jap. *Keman-sō, Yō-raku-botan ;* a perennial herb of the order Papaveraceæ growing in mountains or planted in gardens. In early spring its young plants shoot forth, and in late spring it produces peduncles with many pink flowers disposed in drooping panicles.

757. Gynandropsis viscida, *Bunge.*, Jap. *Fūcho-sō, Yōka-kusa ;* an annual herb of the order Capparidaceæ. In spring it is sown, growing about a foot high, and in autumn its white flowers open at the tops of stems. The flower resemble flying *Fūckō* (the name of a bird), whence the Japanese name is derived.

758. Viola patrinii, *Dc.*, var. chinensis, *Ging.*, Jap. *Sumire, Sumotori-bana ;* a perennial wild herb of the order

Violaceæ. In early spring it shoots forth peduncles and opens dark purple flowers. There are several other varieties, light purple, pink, snow white, etc.

759. Viola pinnata, *L.,* var. dissecta, *Furch.,* Jap. *Yezo-sumire;* a perennial wild herb of the order Violaceæ. In early spring it shoots forth peduncles with leaves, and only one flower blooms on each peduncle. The flower is white with purple stripes and light purple veins.

760. Viola sylvestris, *Kit.,* var. grypoceras, *A. Gray,* Jap. *Tachitsubo-sumire;* a perennial wild herb of the order Violaceæ, growing to a height of about 1 ft. In early summer it bears reddish purple or purplish white flowers.

761. Dianthus superbus, *L.,* Jap. *Nadeshiko, No-nadeshiko, Tokonatsu;* a perennial wild herb of the order Caryophyllaceæ, growing to a height of about 2 fts. In late summer it bears deeply cut thin petaled pink flowers.

762. Dianthus chinensis, *L.,* Jap. *Kara-nadeshiko, Sekichiku;* a biennial garden herb of the order Caryophyllaceæ. It produces many stems from one root, growing to a height of about 6 inches. In summer it blooms at the top of each stem. The flowers are of different colours, as red, white, and variegated.

763. Dianthus chinensis, *L.,* var. hortensis, Jap. *Ise-nadeshiko, Satsuma-nadeshiko, Ōsaka-nadeshiko;* a garden variety of *Nadeshiko* (761) with larger flowers. The flowers are single or double, and pink, purple, or variegated. The petals are cut finely, sometimes drooping 4–5 inches long.

764. Lychnis grandiflora, *Jacq.,* Jap. *Gampi;* a perennial garden herb of the order Caryophyllaceæ. In early spring it sprouts, and in mid-summer it grows to a height of about 3 fts., bearing flowers of various colours. A variety called *Kuruma-gampi* has several opposite leaves, and produces clusters of flowers.

764. b. Lychnis grandiflora, *Jacq.,* var. calicibus,

Jap. *Matsumoto;* a variety resembling the preceding, growing to a height of about 2 fts. In early summer it bears red, white or variegated flowers.

764 c. Lychnis senno, *S.* et *Z.*, Jap. *Sennōke;* a variety resembling the preceding, attaining to a height of about 3 fts. In late summer it bears red, white or other coloured flowers.

765. Lychnis miqueliana, *Rohrb.*, Jap. *Fushiguro-sennō;* a perennial wild herb of the order Caryophyllaceæ. In spring it sprouts, growing in summer to a height of 2-3 fts., when it produces 2 or 3 flowers which are red, white, etc.

766. Silene stellarioides, *Max.*, Jap. *Shirane-gampi;* a perennial herb of the order Caryophyllaceæ growing in high mountains. In spring it grows to height of about 1 ft., and yields white flowers in late summer. It is found in *Shirane*-mountain of Province *Shimo-osa*, whence the name is derived.

767. Silene keiskei, *Miq.*, Jap. *Biranji, Sakura-sennō;* a perennial herb of the order Caryophyllaceæ growing in high mountains. In spring it grows to a height of 4-5 inches, and in late summer reddish purple flowers are produced.

768. Saponaria vaccaria, *L.*, Jap. *Dōkwan-sō;* a biennial garden herb of the order Caryophyllaceæ. It is sown in autumn, and shoots up in the following spring growing to a height of 1-2 fts. In early summer it bears pink flowers on the divided branches.

769. Malva sylvestris, *L.*, Jap. *Zeni-aoi;* a biennial garden herb of the order Malvaceæ. It is sown in autumn, and in the following spring it grows to a height of about 2 fts., when it bears purple, pink or white flowers at the axils of leaves.

770. Althæa rosea, *L.*, Jap. *Tachi-aoi, Hana-aoi;* a garden biennial herb of the order Malvaceæ. It is sown in autumn, and grows to a height of 5-6 fts. in the following spring. It blooms at the axils of leaves in summer. The flowers are single or double, and pink, white or purple.

771. Hibiscus rosa-sinensis, *L.*, Jap. *Riukiu-mukuge*, *Bussōke;* a deciduous shrub of the order Malvaceæ growing in warm provinces, attaining to a height of about 10 fts. In late summer it produces flowers coloured pink, brownish yellow, etc. In winter it is kept in hot houses.

772. Pentapetes phœnicea, *L.*, Jap. *Goji-kwa;* an annual garden herb of the order Sterculiaceæ. It is sown in spring, and in late summer its yellowish red flowers open at noon, whence its name is derived.

773. Hypericum salicifolium, *S.* et *Z.*, Jap. *Byō-yanagi;* a deciduous shrub of the order Hypericaceæ growing wild in mountain-valleys, growing to a height of 4-5 fts. In late summer it bears yellow flowers, and the stamens are very long looking as golden threads.

773. b. Hypericum patulum, *Th.*, Jap. *Kinshibai;* a variety of the preceding, growing to a height of 2-3 fts., but the flower is smaller and the stamens shorter.

774. Hypericum ascyron, *L.*, Jap. *Tomoyesō*, *Byōsō*, *Ō-otogiri;* a perennial wild herb of the order Hypericaceæ, growing to a height of 2-3 fts. In late summer each stalk bears yellowish flowers, which resemble those of the preceding.

775. Geranium eriostemon, *Fisch.*, Jap. *Gunnai-fūro;* a perennial mountain herb of the order Geraniaceæ. In spring it shoots, and in early summer, each peduncle bears light purplish red flowers.

776. Impatiens balsamina, *L.*, Jap. *Hōsenkwa;* an annual garden herb of the order Balsaminaceæ. It is sown in spring, and in summer it attains to a height of about 1 ft. The branches bear single or double, and red, white, purple or variegated flowers.

777. Impatiens textori, *Miq.*, Jap. *Tsurifunesō*, *Hora-gaisō;* an annual herb of the order Balsaminaceæ, growing in shady places as bushes and bamboo woods. It is sown in spring, growing to a height of about a foot in summer. Its flowers resemble the preceding, being single and light reddish purple.

778. Impatiens nolitangera, *Max.*, Jap. *Kitsurifune-sō;* an annual herb of the order Balsaminaceæ growing in shady places of mountains. It grows from the seed sown in spring, and attains to a height of about 2 fts. in summer. It resembles the preceding in form, but the flowers are yellow.

779. Oxalis obtriangulata, *Max.*, Jap. *Yeizan-katabami, Miyama-katabami;* a perennial herb of the order Oxalidaceæ growing in shady places of mountains. Its peduncles shoot up in late spring, opening white flowers with pink veins and light red lines.

780. Bœnninghausenia albiflora, *Reich.*, Jap. *Matsukajesō, Matsugaye-rūda;* a perennial herb of the order Rutaceæ growing in mountains. In summer it grows to a height of 1-2 fts., and produces many small yellowish white flowers.

781. Dictamnus albus, *L.*, Jap. *Hakusen;* a small shrub of the order Rutaceæ planted in gardens. In summer the stems grow to a height of 2-3 fts. Its flowers in panicles are white shaded with purple. The seeds resemble those of Fœniculum vulgare.

782. Thermopsis fabacea, *Dc.*, Jap. *Sendai-hagi;* a perennial herb of the order Leguminoceæ growing principally in the district of *Sendai*, whence its Japanese name is derived. It attains to a height of about 1 ft. It produces yellow papilio-flowers in panicles and flat pods.

783. Crotalaria sessiliflora, *L.*, Jap. *Tanuki-mame;* an annual wild herb of the order Leguminoceæ. It grows from the seed in spring, and in summer it attains to a height of about 1 ft. It produces purple papilio-flowers, being succeeded with hairy pods.

784. Cytisus scoparius, *Link.*, Jap. *Yenishida;* an evergreen garden shrub of the order Leguminoceæ. The stems are dark green, and grow in groups. When several years old, they attain to a height of about 10 fts. In early summer they bear golden yellow papilio-flowers, being succeeded with pods.

785. Spartium junceum, *L.*, Jap. *Redama;* an evergreen shrub of the order Leguminoceæ produced in warm provinces. In winter it is kept in hot houses. The stem is dark green. It blooms in early summer, and the papilio-flowers are yellow and beautiful.

786. Indigofera decora, *Lindl.*, Jap. *Niwa-fuji, Iwa-fuji;* a small wild shrub of the order Loguminoceæ. When young, it looks like a herb. In summer it grows to a height of 1-2 fts., and produces red or white papilio-flowers in panicles.

787. Indigofera tinctoria, *L.*, Jap. *Komatsunagi;* a small wild shrub of the order Leguminoceæ, growing to a height of 1-2 fts. In spring its young branches and leaves shoot forth, bearing in autumn reddish purple or white papilio-flowers.

788. Astragalus sinicus, *Th.*, Jap. *Rengesō, Gengebana, Shōmensō;* a biennial wild herb of the order Leguminoceæ. It creeps along the ground, and produces peduncles with purplish red papilio-flowers arranged in the form of an umbrella. It resembles the lotus flower in shape, though very small, whence its Japanese name is derived. The seeds are sown in autumn, and in the following year the plants are buried under the ground as manure.

789. Lathyrus messerschmidii, *Fr. et Sav.*, Jap. *Nanten-hagi, Tani-watashi, Futaba-hagi;* a perennial wild herb of the order Leguminoceæ. In summer it grows to a height of about a foot, and in autumn it yields reddish purple papilio-flowers, being succeeded with small pods.

790. Lathyrus palustris, *L.*, var. linearifolius, *Ser.*, Jap. *Renrisō, Kamakirisō;* a perennial wild herb of the order

Leguminoceæ. In early summer it grows to a height of about 1 ft., and in autumn it produces peduncles from the axils of leaves, bearing deep purple papilionaceous flowers in panicles.

791. Vicia sativa, *Miq.*, Jap. *Yahadsu-yendō;* a biennial leguminous wild herb. In spring its slender tendrils come forth, and in early summer it produces very small reddish purple papilionaceous flowers at the axils of leaves, and then pods.

792. Lathyrus maritimus, *Miq.*, Jap. *Hama-yendō;* a perennial leguminous herb growing on sandy sea-coasts. In early summer the stem grows and creeps over the ground, producing purple papilionaceous flowers from the axils of leaves.

793. Rubus rosifolius, *Sm.*, var. B. coronarius, *Sims.*, Jap. *Tokin-bara, Tokin-ibara, Goyaogi;* a garden deciduous semi-ligneous shrub of the order Rosaceæ. In summer its slender stem grows like a vine, bearing double yellowish white flowers, which resemble rose-flowers.

794. Lythrum virgatum, *L.*, Jap. *Miso-hagi;* a perennial herb of the order Lythrariaceæ growing in moist places. In summer it grows to a height of about 2 fts. In autumn it produces reddish purple flowers disposed in panicles.

795. Bredia hirsuta, *Bl.*, Jap. *Hashikanboku;* an evergreen small shrub of the order Melastomaceæ growing in warm countries. In autumn it is kept in hot houses. It bears several reddish flowers in panicles at the ends of branches in late autumn.

796. Melastoma macrocarpa, *Don.*, Jap. *Nobotan;* an evergreen shrub of the order Melastomaceæ growing in warm provinces. In water it must be kept in hot houses. In summer it produces reddish purple flowers at the ends of branches and the axils of leaves.

767. Myrtus tomentosa, *Wight*, Jap. *Tenninkwa;* an evergreen shrub of the order Myrtaceæ grown in warm regions. In winter it must be kept in hot houses. In summer it bears pink flowers on small peduncles produced at the axils of leaves and the tops of branches.

798. Epilobium angustifolium, *L.*, Jap. *Yanagi-ran, Yanagi-sō ;* a perennial mountain herb of the order Onagraceæ. In summer it attains to a height of 2–3 fts. Its flowers are reddish purple, being disposed in panicles on the branches. The seeds are provided with fibres.

799. Passiflora cærulea, *L.*, Jap. *Tokei-sō ;* an evergreen climber of the order Passifloraceæ growing in warm regions. In winter it must be kept in hot-houses. In summer it blooms at noon. The flower is provided with many fibrous petals, and the outer petals are white, while the inner petals are purple. Its pistils and stamens resemble a clock in form, whence the Japanese name is derived.

800. Sedum sieboldi, *Sweet*, Jap. *Misebaya-sō, Tamano-o ;* a perennial garden herb of the order Crassulaceæ. It has many drooping stems, and so it is planted in hanging baskets. In summer, it produces small pink flowers in branches at the top of the stem. The plants of this species do not fade without moisture, and thrive well from the cut stems.

801. Sedum erythrostictum, *Miq.*, Jap. *Benkeisō ;* a perennial herb of the order Crassulaceæ planted in gardens. It produces many leaves from one root and grows to a height of about 1 ft. In summer it produces many small pink flowers on the peduncles divided on the top of the stem.

802. Sedum kamtschaticum, *Fisch* et *Mey.*, Jap. *Kirinsō ;* a garden perennial herb of the order Crassulaceæ. It is allied to the preceding, but the leaves are narrow and the yellow flowers are arranged in an umbel.

803. Sedum, Jap. *Iwa-kirinsō ;* a perennial herb of the order Crassulaceæ growing in high mountains. Its stem grows to a height of about 1 ft. and stands obliquely. In late autumn, it opens small yellow flowers disposed in an umbel at the top of the stem.

804. Sedum lineare, *Th.*, Jap. *Mannensō;* a perennial herb of the order Crassulaceæ. It is planted on rocks in gardens. It grows to a height of 6-7 inches, and in summer it yields small 5 petaled golden yellow flowers at the top of the stem.

805. Sedum subtile, *Miq.*, Jap. *Maruba-mannensō;* a variety of the preceding with round leaves. It thrives well by road sides and between rocks. Its stem grows obliquely and reaches to a height of 4-5 inches. In summer it yields small 5 petaled yellow flowers in bunches at the top of the stem.

806. Cotyledon spinosa, *L.*, Jap. *Tsumerenge;* an evergreen herb of the order Crassulaceæ growing on roofs and rocks. It is also planted in pots. In summer its stems grow to a height of 4-6 inches. It opens small pink flowers disposed in panicles.

807. Cotyledon malacophyllum, *Pall.*, Jap. *Iwarenge;* a variety of the preceding, but the leaves are round, broad, and covered with white powder. Its leaf resembles a lotus-flower. It also grows like a lotus.

808. Opuntia fiscus, *L.*, Jap. *Saboten;* a peculiarly formed evergreen shrub of the order Cactaceæ growing in warm provinces. The stem is flat, broad and juicy, and covered with thorns. When young, it has slender leaves. It is about 1 ft. long and 2-3 inches broad and attached one upon another, reaching to a height of about 10 fts. In summer it produces double yellowish red flowers. The fruits resemble figs, and are thorny. They are edible when fully ripe. The young soft stem can be eaten as a vegetable. The juice is used for washing, whence the Japanese name *Saboten* (soap) is derived.

809. Ribes ambiguum, *Max.*, Jap. *Yasha-bishaku, Tembai;* a deciduous small shrub of the order Saxifragaceæ, growing on old trees of high mountains. It grows to a height of

1-2 fts. In summer it blooms, being succeeded with hairy small oval berries edible with a sour taste.

810. **Saxifraga sarmentosa,** *L.*, Jap. *Yukinoshita ;* an evergreen herb of the order Saxifragaceæ growing mountain valleys, and much planted on rock works of gardens. In late summer, it shoots forth long peduncles with white flowers composed of 2 large and 3 small petals.

811. **Saxifraga cortusæfolia,** *S.* et *Z.*, Jap. *Daimojisō, Yukimochisō ;* a perennial herb of the order Saxifragaceæ growing in mountain valleys. In summer it produces white flowers in clusters forming a panicle.

811. b. **Saxifraga cortusæfolia,** *S.* et *Z.*, var. mandida, *Max.*, Jap. *Jinjisō, Kikuba-daimojisō ;* a variety of the preceding with chrysanthemum-like leaves.

812. **Saxifraga sendaica,** *Max.*, Jap. *Sendaisō, Takiwakisō, Harisō ;* a variety of the preceding with its stem 6-7 inches high and thick smooth leaves in clusters. Peduncles shoot forth from the centre and bear small white flowers in bunches resembling the preceding.

813. **Tiarella polyphylla,** *Don.*, Jap. *Dsuda-yakushu ;* a perennial herb of the order Saxifragaceæ growing in mountain-valleys. In summer it bears small white flowers at the top of the stem, being followed with small pods containing fine seeds.

813. b. **Mitella japonica,** *Miq.*, Jap. *Charumerusō, Me-yukinoshita ;* a variety of the preceding. Its flower looks like a trumpet.

814. **Astilbe japonica,** *Miq.*, Jap. *Awamorisō, Awamorishōma ;* a perennial wild herb of the order Saxifragaceæ. The leaves are dark green and lustrous. In summer it grows to a height of about 1 ft., and yields small white flowers in panicles at the top.

815. Rodgersia podophylla, *A. Gray*, Jap. *Yaguruma-sō ;* a perennial herb of the order Saxifragaceæ growing in high mountains. The 5 small leaflets attach to one petiole. In summer it grows to a height of about 2 fts., and bears small white flowers in clusters.

816. Parnassia foliosa, *Hook.*, Jap. *Shirahigesō, Hakusan-nadeshiko ;* a perennial herb of the order Saxifragaceæ growing in high mountains. In summer it shoots forth a peduncle to a height of 6-7 fts. and bears flowers with 5 white fringed petals, resembling those of Dianthus superbus (761).

817. Parnassia palustris, *L.*, Jap. *Mumebachisō ;* a variety closely allied to the preceding growing wild. In late autumn it produces a peduncle of a height of 8-10 inches, and bears yellowish white 5 petaled flowers at the top.

818. Hydrangea stellata, *Max.*, Jap. *Shichidankwa ;* a deciduous shrub of the order Saxifragaceæ growing to a height of 3-4 fts. In summer it blooms light purple flowers. From the centre of the flower it produces another peduncle with flowers, and so on till 5-7 stages.

819. Deinanthe bifida, *Max.*, Jap. *Gingasō, Dangobana, Ginbaisō ;* a perennial herb of the order Saxifragaceæ growing in high mountains, with a height of 1-2 fts. In summer it produces several peduncles with white flowers at the tops, resembling tea-flowers.

820. Hydrangea involucrata, *Sieb.*, Jap. *Tama-ajisai ;* a deciduous semi-ligneous shrub of the order Saxifragaceæ growing in high mountains, with a height of 3-4 fts. In summer it produces globous buds, being followed with small purplish pink flowers in clusters.

821. Hydrangea involucrata, *Sieb.*, var., Jap. *Giokudankwa ;* a variety of the preceding with double flowers. From the centre of the flower it produces another peduncle with flowers. The flowers are greenish white at first, and then turn reddish white.

822. Cardiandra alternifolia, *S.* et *Z.*, Jap. *Kusa-gaku*, *Kusa-ajisai;* a perennial mountain herb of the order Saxifragaceæ, growing to a height of 1½-2 fts. In summer it produces pink flowers in clusters.

823. Fatsia horrida, *Smith*, Jap. *Haribuki*, *Kumadara;* a deciduous mountain shrub of the order Araliaceæ, growing to a height of 4-5 fts. Its leaves and stems are thorny. In summer it shoots peduncles at the top, and yields small light pink flowers in the form of a round bulb.

824. Cornus canadensis, *L.*, Jap. *Gozen-tachibana;* an evergreen herb of the order Cornaceæ growing in shady places of high mountains. It grows to a height of 4-5 inches. From the centre of the 6 leaves at the stem-end, it produces a peduncle with greenish white flowers, being succeeded with edible small red berries.

825. Ixora stricta, *Roxb.*, Jap. *Sandankwa;* an evergreen shrub of the order Rubiaceæ found in warm regions. It does not thrive in cold regions. In summer it produces several peduncles, each with crimson flowers forming an umbel.

826. Gardenia radicans, *Thunb.*, Jap. *Ko-kuchinashi*, *Yaye-kuchinashi;* an evergreen shrub of the order Rubiaceæ, being a variety of Gardenia florida (366). It is much planted in gardens. In summer it produces peduncles with double yellowish white flowers.

827. Damnacanthus indicus, *Gærtn.*, Jap. *Aridōshi*, *Kotori-tomaradsu;* an evergreen small shrub of the order Rubiaceæ found in the mountains of warm provinces, growing to a height of about 2 fts. Its thorny branches come forth in thick bushes. In early summer, it bears clove-like small white flowers, being succeeded with small round red berries which remain on the branches till the new berries of the next year are produced.

828. Patrinia scabiosæfolia, *Link.*, Jap. *Ominayeshi*, *Awa-bana;* a perennial wild herb of the order Valerianaceæ, growing to a height of 3-4 fts. In late summer it produces beautiful yellow flowers in an umbel at the top of the stem.

829. Scabiosa japonica, *Miq.*, Jap. *Matsumushisō, Rimbō-giku ;* a wild biennial plant of the order Dipsaceæ, growing to a height of 2-3 fts. In autumn it shoots forth peduncles with small purple flowers in a composite form. Its young leaves are edible.

830. Eupatorium chinense, *Miq.*, Jap. *Fuji-bakama ;* a wild perennial herb of the order Compositæ, growing to a height of 3-4 fts. Late in autumn, it bears fragrant purple flowers at the top.

831. Aster trinervius, *Roxb.*, var. congesta, *Fr.* et *Sav.*, Jap. *Kon-giku ;* a wild perennial herb of the order Compositæ closely allied to Aster cantoniensis (65). It grows to a height of about 1 ft., and in late autumn it bears deep purple flowers.

832. Aster spathulifolius, *Max.*, Jap. *Daruma-giku, Shinano-giku ;* a garden perennial herb of the order Compositæ, growing to a height of about 1 ft. Its leaves are covered with fine hair. In autumn it produces several purplish pink flowers at the top.

833. Aster tataricus, *L.*, Jap. *Shion ;* a garden perennial herb of the order Compositæ, growing straight 5-6 fts. high. In late autumn it bears purplish flowers. There is a dwarf variety, being about 1 ft. high.

834. Aster cantoniensis, *Dc.*, var., Jap. *No-shungiku, Shungiku ;* a garden perennial herb of the order Compositæ, growing to a height of about 1 ft. From early summer to autumn it bears reddish purple flowers. There is also a variety with white flowers.

835. Solidago virga-aurea, *L.*, Jap. *Akino-kirinsō, Awadachisō ;* a perennial wild herb of the order Compositæ, growing to a height of 1-2 fts. In late autumn it bears 5 petaled small yellow flowers in panicles. There is also a variety with white flowers.

836. Pyrethrum, Jap. *Ō-giku;* a perennial garden herb of the order Compositæ. There are two varieties, the summer and the autumn crysanthemums. The one here mentioned is the autumn variety. It grows to a height of 2–4 fts., and in early autumn it opens flowers with diverse colours, red, yellow, white, orange, etc. Some of the flowers are several inches in diameter. Their petals are also various, flat, tubular, etc. They are the best of the autumnal flowers.

837. Pyrethrum, Jap. *Natsu-giku;* a perennial garden herb of the order Compositæ, growing to a height of 2–3 fts. In summer it bears flowers of several colours and forms.

838. Pyrethrum, Jap. *Ko-giku;* a perennial garden herb of the order Compositæ, growing to a height of about 1 ft. In late autumn its divided branches bear flowers of yellow, white, or red colours and of different sizes. They are mostly derived from *Iwa-giku*.

838. b. Pyrethrum, Jap. *Kan-giku;* a variety of the preceding, bearing small yellowish flowers in late autumn.

839. Leucanthemum nipponicum, *Fr.* et *Sav.*, Jap. *Hama-giku;* a perennial herb of the order Compositæ, growing wild on sea-coasts and also planted in gardens. Its stems do not die through the year, growing about 2 fts. high. In late autumn it bears white flowers with a yellow centre at the tops of the branches.

840. Leucanthemum arcticum, *Dc.*, Jap. *Ko-hama-giku;* a perennial herb of the order Compositæ, growing on the sea-coasts of northern provinces. It grows to a height of about 1 ft., and in late autumn it produces white flowers which turn purplish pink when old.

841. Callistephus chinensis, *Nees.*, Jap. *Ezo-giku, Satsuma-kon-giku;* a biennial garden plant of the order Compositæ. It becomes also an annual plant according to the season in which the seeds are sown. It grows to a height of 1-2 fts. The biennial one blooms in summer, and the annual one in autumn. The flowers are purplish blue, red, white, etc.

842. Achillea sibirica, *Led.*, Jap. *Hagoromosō, Nokogirisō;* a perennial wild herb of the order Compositæ, growing to a height of 2-3 fts. In autumn it produces small white or pink flowers in clusters at the head.

843. Artemisia schmidtiana, *Max.*, Jap. *Asagirisō;* an evergreen herb of the order Compositæ produced in northern provinces. Its leaves are slender, green, and lustrous. It grows to a height of 1-2 fts., and in autumn it bears small yellow flowers in panicles.

844. Gnaphalis japonica, *Max.*, Jap. *Arare-giku, Yama-hahako;* a perennial wild herb of the order Compositæ, growing to a height of about 1 ft. Late in autumn, it bears many small white flowers with yellow centres. The flowers remain still after the stems were withered by frost.

845. Senecio japonica, *Schultz.*, Jap. *Hankwaisō;* a perennial wild herb of the order Compositæ. Its leaves are broad and deeply dissected. It grows to a height of about 3 fts., and in autumn it blooms yellow flowers on the branches divided at the head of the stem. There is a variety called *Chōriōsō* closely allied, but the dissection of the leaves is less and it grows to a height of about 5 fts.

846. Senecio kæmpferi, *Dc.*, Jap. *Tsuwabuki;* an evergreen herb of the order Compositæ, growing wild on the seacoasts of southern provinces and also much planted in gardens. In autumn its stems grow to a height of about 2 fts. and divided into branches, bearing yellow flowers. The petioles of the young leaves are eaten as a vegetable (67. b.). The variety called *Ō-tsuwabuki* is large, and the variety called *Kan-tsuwabuki* blooms in winter.

847. Senecio flammeus, *Dc.*, Jap. *Kōrinkwa;* a perennial herb of the order Compositæ closely allied to Senecio campestris, growing wild in the dry places of mountains and fields. It grows to a height of about 1 ft., yielding many reddish yellow flowers at the top of the divided branches.

848. Calendula officinalis, *L.*, Jap. *Kinsenkwa;* a biennial garden herb of the order Compositæ. It becomes annual according to the time of its sowing. It grows to a height of about 1 ft. The biennial blooms in late spring, and the annual in late autumn. The flowers are reddish yellow or light yellow, and a variety with large flowers is called *Tōkinsen*.

849. Echinops sphærocephalus, *L.*, Jap. *Ruri-higodai, Higodai;* a perennial wild herb of the order Compositæ, growing to a height of 2-3 fts. In autumn it bears composite flowers forming very pretty purplish blue balls at the top of the divided stems.

850. Cnicus spicatus, *Max.*, Jap. *Yama-azami, Oniazami;* a perennial wild herb of the order Compositæ growing to a height of about 2 fts. The leaves and stems are thorny. In late summer, it is divided into branches at the top, and bears reddish purple flowers.

850. b. Cnicus buergeri, *Max.*, Jap. *No-azami;* it resembles the preceding, but smaller. It blooms in early summer, and the flowers are purple. There is a variety called *Hana-azami* with pretty flowers of diverse colours, red, white, etc.

851. Rhaponticum atriplicifolium, *Dc.*, Jap. *Kumatori-bokuchi, Yama-gobō, Yama-hokuchi;* a perennial wild herb of the order Compositæ, growing to a height of 3-4 fts. Its stems and leaves are covered with fine white hair. In autumn it produces globular thorny buds, and then dark purple flowers. The young leaves are edible, and the old leaves are used to make a tinder.

852. Serratula coronata, *L.*, Jap. *Tamurasō, Tamabōki;* a perennial wild herb of the order Compositæ, growing to a height of 2-3 fts. In autumn it bears reddish purple flowers at the top of the branches. It resembles Cnicus spicatus, but has no thorn.

853. **Taraxacum officinale,** *Wigg.,* Jap. *Tampopo ;* a perennial herb of the order Compositæ growing wild everywhere. In spring it shoots peduncles among the leaves, and yields deep yellow flowers at the top. There are several varieties of various flowers and leaves. The young leaves are eaten as a vegetable, being soft and delicious.

854. **Helianthus annuus,** *L.,* Jap. *Himawari, Hi-guruma, Nichirinsō ;* an annual garden plant of the order Compositæ, growing to a height of 6-7 fts. In autumn it bears one yellow flower at the head of each stem. The flower is 8-9 inches in diameter, and turns round towards the sun. The seeds are used to take an oil.

855. **Platycodon grandiflorum,** *Dc.,* Jap. *Kikyō ;* a perennial wild herb of the order Campanulaceæ, growing to a height of 2-3 fts. In autumn it shoots peduncles at the top, and bears purplish blue flowers. There are many varieties planted in gardens.

856. **Wahlenbergia marginata,** *Dc.,* Jap. *Hina-gikyō ;* a perennial wild herb of the order Campanulaceæ. Its slender stem grows obliquely to a height of 4-5 inches. In late summer, the stem shoots branches and bears small bluish purple flowers.

857. **Glossocomia lanceolata,** *Reg.,* Jap. *Tsuru-ninjin ;* a perennial wild climber of the order Campanulaceæ. In autumn it produces flowers from the axils of leaves. The flower is greenish white with dark purple vein-nets in the outside, and dark purple vein-nets and spots in the inside. A variety called *Basobu* is covered with fine hair, and its leaves are thin.

858. **Phyteuma japonicum,** *Miq.,* Jap. *Shide-shajin. Chiji-gikyō ;* a perennial wild herb of the order Campanulaceæ,

growing to a height of 2-3 fts. In late summer it bears narrow petaled purple or white flowers in panicles at the head of the stem.

859. Campanula punctata, *Lamk.*, Jap. *Hotaru-bukuro, Tsuriganesō, Chōchin-bana;* a perennial wild herb of the order Campanulaceæ, growing to a height of 2-3 fts. In summer it bears campanulate purplish flowers with deep purple spots or white flowers with purple spots. Those growing on high mountains have a height of 3-4 inches, and their flowers are very pretty.

860. Adenophora trachelioides, *Max.*, Jap. *Sobana;* a perennial herb of the order Campanulaceæ growing in mountains. In late summer it grows to a height of 2-3 fts., and bears light purple campanulate flowers in panicles.

861. Adenophora verticillata, *Fisch.*, Jap. *Tsurigane-ninjin;* a perennial wild herb of the order Campanulaceæ, growing to a height of 3-4 fts. In summer it produces small bluish purple or white campanulate flowers in panicles.

862. Adenophora denticulata, *Th.*, Jap. *Hime-shajin;* a small variety of the preceding growing on high mountains, growing to a height of about 2 fts. In autumn it produces greenish purple campanulate flowers in panicles at the head of the stem.

863. Campanula glomerata, *L.*, var. genuina, *Herd.*, Jap. *Yatsushiro-gikyō;* a perennial herb of the order Campanulaceæ growing in moist places. In late summer it grows to a height of about 2 fts., and yields campanulate purplish blue flowers in clusters from the axils of leaves at the head of the stem.

864. Conandron ramondioides, *S. et Z.*, Jap. *Iwa-na, Iwa-jisha, Iwa-tabako;* a perennial herb of the order Cyrtandraceæ growing on rocky mountains, producing one leaf from each root. Its stalk grows in summer, divided into branches, with several flowers which are pink, purple, or white. In late spring its young leaves are eaten as vegetables.

865. Rehmannia glutinosa, *Libosch.*, Jap. *Senrigoma;*

a perennial garden herb of the order Cyrtandraceæ. In summer it grows to a height of about 1 ft. and bears labiated flowers. The out-side of the flower is light red shaded with purple, and the inside is yellow with purple spots.

866. Andromeda polifolia, *L.*, Jap. *Hime-shakunage;* an evergreen small shrub of the order Ericaceæ growing in moist places of high mountains. In summer it grows to a height of about 1 ft. and produces peduncles at the top of the branches, bearing one campanulate light red flower on each peduncle.

867. Phyllodoce taxifolia, *Don.*, Jap. *Tsuga-zakura;* an evergreen small shrub of the order Ericaceæ growing on high mountains, growing to a height of about 1 ft. In summer it bears small light red campanula with five petals.

868. Pyrola rotundifolia, *L.*, Jap. *Ichiyakusō, Kikko-sō;* an evergreen wild herb of the order Ericaceæ. In summer it shoots peduncles to a height of about 8-9 inches, and bears yellowish white flowers. There are several varieties.

869. Chimaphila japonica, *Max.*, Jap. *Mumegasasō, Kinugasasō;* a small evergreen herb resembling the preceding, growing in shady places of mountains. In summer it shoots 6-8 inches long peduncles from the axils of leaves, and bears greenish white flowers in panicles.

870. Chloranthus brachystachys, *Bl.*, Jap. *Senryō;* an evergreen herb of the order Chloranthaceæ produced in warm provinces, growing to a height of 2-3 fts. In summer it produces peduncles at the top of the branches, and bears small yellowish green flowers in clusters, being succeeded with small round red berries. The fruits remain on the branches till the following year. There is a variety with white berries.

871. Ardisia crispa, *Dc.*, Jap. *Manryō;* an evergreen shrub of the order Myrsinaceæ growing in shady places of mountains. It grows to a height of 2-3 fts., but 7-8 fts. in warm regions. In summer it produces peduncles, and bears small white flowers, being succeeded with round red berries which remain on the branches for a long time. There are varieties with white or yellow berries.

872. Bladhia crenata, Jap. *Karatachi-bana, Kōji;* an evergreen shrub of the order Myrsinaceæ produced in warm regions, growing to a height of about 1 ft. In summer it bears greenish white flowers, being succeeded with small round berries, which are red, yellow, or white. The leaves are various.

873. Ardisia japonica, *Bl.,* Jap. *Yabu-kōji, Yabu-tachi-bana;* an evergreen wild shrub of the order Myrsinaceæ, growing to a height of about 1 ft. In summer it produces bluish white flowers at the axils of leaves, being succeeded with small round red berries. There is a variety with white berries which remain for a long time on the branches. There are also several varieties with various leaves.

874. Asclepias curassavica, *L.,* Jap. *Tōwata;* an annual garden herb of the order Asclepiadaceæ growing to a height of 2-3 fts. In summer it produces red flowers at the top, being succeeded with pods. When ripe, the pods open and expose the seeds provided with white fibrous tufts.

875. Hoya carnosa, *R. Br.,* Jap. *Sakura-ran;* an evergreen climber of the order Asclepiadaceæ produced in warm provinces In winter it must be kept in hot-houses. In summer it produces peduncles bearing light red bell-flowers.

876. Amsonia elliptica, *Ræm.* et *Sch.,* Jap. *Chōji-sō;* a perennial wild herb of the order Apocynaceæ growing to a height of 2-3 fts. In summer it produces greenish purple flowers. Its ripe pods contain seeds provided with fibrous tufts.

877. Nerium odorum, *Soland,* Jap. *Kiōchikutō;* an evergreen shrub of the order Apocynaceæ found in warm regions growing to a height of about 1 ft. In late summer it bears pink, purple, or white flowers.

878. **Vinca rosea**, *L.*, Jap. *Nichinichisō, Nichinichikwa;* an annual herb of the order Apocynaceæ brought formerly from foreign countries, growing to a height of about 1 ft. In late summer it blooms every day. The flowers are pinkish purple or white.

879. **Gentiana scabra**, *Bunge*, var. Buergeri, *Max.*, Jap. *Rindō, Sasa-rindō;* a perennial wild herb of the order Gentianaceæ, growing to a height of 1-2 fts. In late autumn it bears blue flowers in clusters. The bitter roots are used for medicine.

880. **Gentiana squarrosa**, *Ledeb.*, Jap. *Koke-rindō, Haru-rindō;* an annual wild herb of the order Gentianaceæ, growing to a height of 4-5 inches. In spring it produces purplish brown flowers. A variety called *Fude-sō* is large in form.

881. **Crawfurdia japonica**, *S. et Z.*, Jap. *Tsuru-rindō;* a perennial herb of the order Gentianaceæ growing in shady places of mountains. The stem is slender, creeping on the ground like a climber. In autumn it produces 3-4 purple flowers at the axils of leaves, being succeeded with oval red berries.

882. **Menianthes trifoliata**, *L.*, Jap. *Mitsu-gashiwa;* a perennial herb of the order Gentianaceæ growing in shallow water. The leaves are ternate. In summer it produces peduncles bearing several light red 5-parted hairy flowers.

883. **Calystegia japonica**, *Miq.*, Jap *Hirugao;* a perennial climbing herb of the order Convolvulaceæ growing everywhere. In summer it produces funnel-shaped flowers at the axils of leaves. The roots are edible when boiled.

884. **Calystegia japonica**, *Miq.*, var. integrifolia, *Fr. et Sav.*, Jap. *Ōhirugao;* a large variety of the preceding, with large pretty flowers.

885. **Calystegia soldanella**, *R. Br.*, Jap. *Hama-hirugao, Aoi-kadsura;* a perennial creeper of the order Convolvulaceæ

growing in sandy places near sea-coasts. In late summer it bears red flowers resembling *Hirugao* in form.

886. Ipomæa bona-box, *L.*, Jap. *Hari-asagao, Chūji-asagao;* an annual climbing herb of the order Convolvulaceæ, brought from foreign lands. The vine in covered with soft pricks. In late summer it produces purple funnel-shaped flowers. The receptacles of the seeds are big and droop by their own weight. The young fruits are eaten as a vegetable.

887. Pharbitis triloba, *Miq.*, Jap. *Asagao;* an annual turning herb of the order Convolvulaceæ planted in gardens. In late summer it blooms at the leaf-axils only in early morning. The flowers and leaves are various.

888. Omphalodes krameri, *Fr.* et *Sav.*, Jap. *Rurisō;* a perennial wild herb of the order Boraginaceæ, growing to a height of 8–9 inches. In summer it bears several blue flowers at the top of its stem. A variety with red flowers is called *Sangosō*.

889. Omphalodes krameri, *Fr.* et *Sav.*, var., Jap. *Hari-sō;* a variety of the preceding with white flowers.

890. Veronica longifolia, *L.*, var. japonica, *Max.*, Jap. *Ruri-torano-o;* a perennial wild herb of the order Scrophulariaceæ, growing to a height of 1–2 fts. In late summer it produces small 4-petaled purplish blue flowers in panicles.

891. Veronica incana, *L.*, Jap. *Tōtci-ran, Hama-tora-no-o;* a perennial herb of the order Scrophulariaceæ growing to a height of about 2 fts. The stems and leaves are hairy. In late summer it bears small pinkish purple flowers in panicles.

892. Veronica onœi, *Fr.* et *Sav.*, Jap. *Hiyokusō;* a perennial wild herb of the order Scrophulariaceæ, growing to a height of about 1 ft. It produces 2 branches at each leaf-axil. In summer it bears purple flowers in panicles.

893. Veronica sibirica, *L.*, Jap. *Kukaisō;* a perennial wild herb of the order Scrophulariaceæ growing to a height of 3-4 fts. Several leaves grow at the same point, forming layers. In summer it bears small purplish blue flowers in panicles.

894. Pæderota villosula, *Miq.*, Jap. *Sudsukakeso* ; a perennial wild herb of the order Scrophulariaceæ. Its stem is slender and creeps over the ground. In summer it produces small purplish blue flowers forming a ball.

895. Pedicularis gloriosa, *Biss.* et *Moor.*, Jap. *Hankwai-azami* ; a perennial herb of the order Scrophulariaceæ found in shady places of mountains. The stem grows to a height of 2-3 fts. In late summer it bears purplish pink labiate flowers in panicles at the top of the stem.

896. Linaria japonica, *Miq.*, Jap. *Kingioso, Unran* ; a perennial herb of the order Scrophulariaceæ growing in sandy places near sea-coasts. The stem grows obliquely about 7-8 inches high, and is covered with white powder. In summer it bears yellowish white labiate flowers at the top of the stem.

897. Pedicularis resupinata, *L.*, Jap. *Shiogama-giku, Shiogama-so* ; a perennial mountain herb of the order Scrophulariaceæ, growing to a height of 1-2 fts. In mid-autumn it produces pinkish purple or yellowish white labiate flowers in clusters at the top of the stem.

898. Nepeta subsessilis, *Max.*, Jap. *Misogawaso* ; a perennial herb of the order Labiatæ growing in moist places of mountains. It grows to a height of 2-3 fts. In early autumn it produces peduncles from the axils of leaves at the top, and bears purple labiate flowers in clusters.

899. Plectranthus longitubus, *Miq.*, Jap. *Kiritsubo, Aki-choji* ; a perennial mountain herb of the order Labiatæ, growing to a height of 2-3 fts. In autumn it shoots peduncles bearing a purplish blue labiate flower on each peduncle.

900. Primula cortusoides, *L.*, Jap. *Sakuraso, Kuruma-*

bana; a perennial wild herb of the order Primulaceæ. In spring it shoots a stem amidst dense leaves to a height of 7-8 inches, and the stem is divided into several peduncles, yielding pinkish purple flowers which resemble cherry-flowers. Varieties planted in gardens are numerous.

901. **Primula japonica,** *A. Gray*, Jap. *Kurinsō;* a perennial mountain herb of the order Primulaceæ. In spring its stalk shoots forth in the dense leaves to a height of about 1 ft., bearing crimson flowers in circles and layers. Those planted in gardens have flowers of different colours, crimson, pink, variegated, white, etc.

902. **Primula kisoana,** *Miq.*, Jap. *Kakkosō;* a perennial herb of the order Primulaceæ growing wild in mountain-valleys. In late spring it produces a stalk with several flowers at the top. The flowers are pinkish purple, but some are white with red stripes.

903. **Gomphrena globosa,** *L.*, Jap. *Sennichisō;* an annual garden herb of the order Amaranthaceæ. It is sown in spring, and grows to a height of 1-2 fts., dividing into branches. It blooms in a globular form, and the flowers are red, white, etc.

904. **Celosia argentea,** *L.*, Jap. *Keitō;* an annual garden herb of the order Amaranthaceæ. It is sown in spring, and grow to a height of 2-3 fts. in summer. Its peduncle covered with small red flowers resembles a cox-comb. The leaves are also red. There are several varieties.

905. **Polygonum cuspidatum,** *S. et Z.*, Jap. *Itadori;* a perennial wild herb of the order Polygonaceæ, growing to a height of 7-8 fts. Late in summer, it produces greenish white small flowers in panicles at the axils of leaves. Its young stalks are eaten as a vegetable (75. b.). A variety with small leaves and pink flowers is called *Meigetsusō.*

906. **Polygonum blumei,** *Meisn.*, Jap. *Sakura-tade, Tademodoki;* an annual herb of the order Polygonaceæ growing

near water. It grows to a height of about 2 fts., and in autumn it produces panicles at the top and the axils of leaves. Its flowers are pink, crimson or white.

907. Polygonum orientale, *L.*, var. pilosum, *Meisn.*, Jap. Ōke-tade, Hotaru-tade ; an annual wild or garden herb of the order Polygonaceæ, growing to a height of 5–6 fts. In autumn it produces panicles at the top and the axils of leaves, with many red flowers in clusters.

908. Polygonum bistorta, *L.*, Jap. Ibuki-torano-o ; a perennial mountain herb of the order Polygonaceæ. In summer its stalk shoots forth to a height of 1–2 fts., and bears pink or white flowers in panicles at the top of the stem.

909. Polygonum filiforme, *Th.*, Jap. Midsuhiki ; a perennial herb of the order Polygonaceæ growing in forests and bushes. It grows to a height of about 2 fts., and in summer it produces long filiform red panicles at the end of the branches.

910. Begonia evansiana, *Andr.*, Jap. Shū-kaidō ; a perennial garden herb of the order Begoniaceæ. It has male and female flowers. It thrives well in moist shady places. It grows to a height of about 2 fts.. and in autumn it produces red flowers at the axils of leaves. The leaves are irregular heart-shaped. This plant is juicy and aciduous.

911. Asarum blumei, *Duch.*, Jap. Kan-aoi ; an evergreen herb of the order Aristolochiaceæ growing in shady places of high mountains. In winter it blooms near the roots and the flowers resemble those of Asarum sieboldi (456), but are yellowish green. It is prized for its pretty variegated leaves.

912. Pachysandra terminalis, *S. et Z.*, Jap. Kichijisō, Fukkisō ; an evergreen mountain-herb of the order Euphorbiaceæ, growing to a height of about 1 ft. In late summer it produces panicles at the top, bearing small yellowish green flowers, which ar esucceeded with small round white berries.

913. Cymbidium, Jap. *Me-ran;* an evergreen herb of the order Orchideæ resembling *Suruga-ran* (895) in form, though with broad leaves and tender nature. It is admired as a pot-plant·

914. Cymbidium virens, *L.*, Jap. *Hokuro, Hakuri, Kusa-ran;* an evergreen wild orchid. In spring it shoots forth peduncles with yellowish green little fragrant flowers. The flowers are edible when preserved in salt.

915. Calanthe japonica, *Bl.*, Jap. *Karan, Riukiu-yebine, Kwaran;* a perennial orchid produced in warm provinces. In cold regions it must be kept in hot-houses during winter. In late summer it produces peduncles to a height of about 1 ft., and bears red, purple or white flowers.

916. Bletia hyacinthina, *R. Br.*, Jap. *Shiran, Shuran, Shikei;* a perennial orchid. Late in spring, it shoots forth peduncles to a height of about 1 ft., and bears 5–6 reddish purple or white flowers at the top. This plant thrives well in gardens. The roots are used to make paste.

917. Phajus maculatus, Jap. *Ganzeki-ran, Ishi-ran;* an evergreen orchid produced in warm provinces. Late in summer, it shoots forth its peduncle to a height of about 1 ft., with several golden yellow flowers. A variety with yellow variegation on the leaves is called *Hoshikei*, and when its variegation is very fine it is called *Kinkei*.

918. Anglæcum falcatum, *B. et H.*, Jap. *Fū-ran;* an evergreen orchid growing on the old tree-trunks of high mountains in warm regions. In summer it produces a peduncle to a height of 5–6 inches, and the peduncle is divided into branches, with slightly fragrant white flowers.

919. Cymbidium japonicum, *Miq.*, Jap. *Nagi-ran;* an evergreen orchid produced in warm provinces. In late spring and early summer, it shoots a peduncle to a height of 6–7 inches, bearing several yellowish white flowers.

920. Aerides japonicum, *Lindl.*, Jap. *Nago-ran;* an evergreen orchid growing on old trees in mountains of warm regions. In summer it produces peduncles 6–7 inches high, and bears several reddish white slightly fragrant flowers.

921. Cleisostoma ionosmum, *Lindl.*, Jap. *Niumen-ran;* an evergreen orchid growing on *Irimote* mountain of *Okinawa* Islands. The stem attains to a height of 2–3 fts. In summer it shoots peduncles, and yields several yellow flowers with reddish brown spots.

922. Cypripedium japonicum, *Th.*, Jap. *Kumagayesō, Hoteisō;* a perennial wild orchid, growing to a height of about 1 ft., with 2 large broad leaves. From spring to summer, it blooms purse-like yellowish green flowers with dark purple spots. There is a variety called *Atsumorisō* (C. macranthum, *Sw.*).

923. Habenaria radiata, *Th.*, Jap. *Sagi-sō;* a perennial orchid growing in marshy places. In spring it produces its peduncle about 1 ft. high, and yields 2–3 white fringed flowers.

924. Goodyera schlechtendaliana, *Reich.*, Jap. *Kamome-ran, Miyama-udsura, Toyoshima-ran;* an evergreen wild orchid, growing to a height of 4–5 inches, with white variegated leaves. From summer to autumn, it bears reddish white flowers in panicles.

925. Dendrobium moniliforme, *Sw.*, Jap. *Sekkoku, Iwadokusa;* an evergreen orchid growing on rocks or old trees of mountains, attaining to a height of several inches. The stems have joints like Equisetum hyemale. In summer it produces light pink flowers. There are several varieties.

926. Luisia teres, *Bl.*, Jap. *Bō-ran, Matsu-ran;* an evergreen orchid, growing on old trees in warm regions about 1 ft. high. In summer it bears yellowish green flowers with dark red spots.

927. Canna indica, *L.*, Jap. *Dandoku;* an evergreen-herb of the order Marantaceæ growing to a height of 2-3 fts. In cold regions, the plant fades, but the root remains like a perennial. In summer it opens several orange-red flowers. The ripe seeds are round, black, and hard, and they sprout readily when sown.

928. Musa coccinea, *Willd.*, Jap. *Hime-bashō, Bijinshō;* an evergreen herb of the order Musaceæ produced in warm regions. It does not thrive in cold weather. It grows to a height of 4-5 fts., and in autumn it yields deep red flowers in layers.

929. Iris tectorum, *Max.*, Jap. *Ichihatsu;* a perennial herb of the order Iridaceæ planted in gardens and sometimes on straw-roofs. In spring it shoots forth its stalks to a height of about 1½ fts., and in summer it bears several purplish green or white flowers.

930. Iris lævigata, *Fisch.*, Jap. *Kakitsubata;* a perennial herb of the order Iridaceæ planted in shallow water growing to a height of 2-3 fts. In summer it produces purplish blue, white, red, or blue flowers.

931. Iris lævigata, *Fisch.*, var. kæmpfereri, Jap. *Hana-shōbu;* a variety of the preceding, blooming earlier. The flowers are purplish blue, white, or variegated, and very pretty.

932. Iris sibirica, *L.*, var. orientalis, Jap. *Ayame, Hana-ayame;* it grows wild in marshy places, attaining to a height of about 1 ft. In early summer it bears purplish blue or white flowers.

933. Iris ensata, *Th.*, var. chinensis, Jap. *Neji-ayame;* it is closely allied to the preceding, but the leaves are twisted. It blooms in early summer, and the flowers are white with purple stripes. Its fibrous roots are fine and strong (348. b.).

934. Iris sibirica, *L.*, var. hæmatophylla, Jap. *Kama-yama-shōbu;* a perennial garden herb of the order Iridaceæ, resembling *Ayame* (932), with leaves 2-3 fts. long. As the leaves are strong and flexible, they are used for tying.

935. **Iris japonica,** *Th.*, Jap. *Shaga;* an evergreen shrub of the order Iridaceæ, growing in shady places. It shoots forth its leaves obliquely. In summer its stems grow to a height of about 1 ft., and produces several purple shaded white flowers with yellow centres.

936. **Iris gracilipes,** *A. Gray*, Jap. *Hime-shaga;* a perennial garden herb of the order Iridaceæ. It resembles the former in form, but smaller. In early summer it shoots forth its stems to a height of about 1 ft., and produces several purple or white flowers.

937. **Iris,** Jap. *Kin-kakitsu, Ko-kakitsubata;* a perennial garden herb of the order Iridaceæ. From spring to summer, it shoots forth straight flower stalks to a height of 5-6 inches, and yields 2-3 golden yellow flowers.

938. **Pardanthum chinensis,** *A. Gray*, Jap. *Hi-ōgi;* a perennial wild herb of the order Iridaceæ. Its leaves grow straight, and from summer to autumn it shoots forth peduncles to a height of 2-3 fts. The colours of the flowers differ according to their own varieties, as red, yellow, etc., and also the leaves are various, short, twisted, etc.

939. **Narcissus tazetta,** *L.*, var. chinensis, *Boem.*, Jap. *Suisen, Gindai;* a bulbous plant of the order Amaryllideæ growing on sea-coasts of warm provinces. In winter it shoots up its peduncle to a height of about 1 ft. from the centre of the slender leaves and bears several flowers out of the bracts. The flowers are 6-petaled and light yellow with funnel shaped coronets. The bulbs grow well in water, and they are very poisonous, but the fresh bulbs are used as medicine.

940. **Narcissus tazetta,** *L.*, var. fl. pleno., Jap. *Yaye-suisen;* the double variety of the preceding. There are several varieties with narrow leaves, narrow petals, or green flowers. They are all precious winter-blooming plants.

941. **Crinum asiaticum,** *L.*, var. declinatum, *Kunth.*,

Jap. *Hama-yū, Hama-omoto ;* an evergreen bulbous plant of the order Amaryllidaceæ, growing on sea-coasts of warm regions. When fully grown, it attains to a height of 4–5 fts., with several large broad leaves in the upper part. In summer it shoots forth peduncles in the centre of leaves, and blooms about ten flowers in an umbel form. The flowers are 6-petaled and white.

941. b. Nerine japonica, *Miq.,* Jap. *Higan-bana ;* this bulbous plant (509) grows abundantly everywhere, and so not precious, but its red flowers are beautiful.

941. c. Lycoris radiata, *Herb.,* Jap., *Kitsune-no-kamisori ;* this bulbous plant (510) has also beautiful flowers.

941. d. Nerine sarniensis, *L.,* Jap. *Shōki-ran ;* a variety of the preceding, with broad leaves and pretty yellow flowers.

941. e. Amaryllis squemigera, *Max.,* Jap. *Natsudsuisen ;* a variety closely allied to the preceding, with reddish purple flowers.

942. Lilium krameri, *Th.,* Jap. *Sasa-yuri,* Yama-yuri;* a wild bulbous plant of the order Liliaceæ. In summer its stalks grow to a hight of 2–3 fts., and bear 6-petaled reddish whith fragrant flowers at the top. The bulbs are edible (121. b).

942. b. Lilium tigrinum, *Gawl.,* Jap. *Oni-yuri ;* this (121) is principally noted for its edfbie roots, but the flowers are also pretty. The varieties with double or yellow flowers or with flat peduncles are especially beautiful.

942. c. Lilium callosum, *S. et Z.,* Jap. *Suge-yuri ;* a wild variety of the preceding with slender leaves.

943. Lilium japonicum, var., Jap. *Satsuki-yuri, Sa-yuri ;* a variety of the preceding, blooming earlier, with red pollen.

944. Lilium auratum, *Lindl.,* Jap. *Hōraiji-yuri ;* its flowers are white with a yellowish band and dark red spots. It is produced abundantly in *Horaiji*-mountain of Province *Mikawa*, whence the Japanese name is derived.

945. Lilium auratum, *Th.*, var. rubro-vittatum, Jap. *Beni-suji-yuri;* a variety of *Sasa-yuri* (942). The flowers have a red band and dark red spots. It is mostly propagated by cultivation.

946. Lilium speciosum, *Th.*, var. rubrum, Jap. *Kano-ko-yuri;* it is cultivated in gardens, growing to a height of 2-3 fts. In summer, it bears several flowers at the top of its stem. The flowers are recurved like those of L. tigrinum (121), and are pink-shaded white with many scarlet spots. The bulbs are yellow and bitter, being not edible.

946. b. Lilium speciosum, *Th.*, var. album, Jap. *Shiro-kanoko-yuri, Shiratama-yuri, Mine-no-yuki;* a variety of the preceding with white flowers. Its bulbs are less bitter and edible.

947. Lilium hansoni, *Leichtl.*, Jap. *Takeshima-yuri;* a species of lilies growing to a height of 2-3 fts. Several leaves grow together and form layers. In summer it produces under-recurved and purple-spotted red flowers.

948. Lilium batemanni, Jap. *Hirato-yuri;* it grows to a height of 2-3 fts. In summer it bears several flowers at the top. The flowers are orange-red, yellow or red.

949. Lilium concolor, *Salisb.*, var. pulchellum, *Fisch.*, Jap. *Hime-yuri, Aka-hime-yuri;* a smallest garden lily, growing to a height of about 1 ft. In summer it bears several red flower at the top.

949. b. Lilium concolor, *Salisb.*, Subr. Coridon, Jap. *Ki-hime-yuri;* a variety of the preceding, with yellow flowers.

950. Lilium thunbergianum, *Roem. et Schult.*, Jap. *Natsu-sukashi-yuri;* it grows on sea-coasts of southern provinces.

It is about 1 foot high, with several red flowers facing upwards. There is an empty space between each petal. It is admired as a flower-plant, and the bulbs are edible (121. c).

950. b. Lilium thunbergianum, *K.* et *S.*, var., Jap. *Haru-sukashi-yuri ;* it blooms early, and the flowers are various.

951. Lilium longiflorum, *Th.*, Jap. *Teppō-yuri, Riu-kiu-yuri ;* it is planted in gardens growing to a height of 1-2 fts. In summer it opens many flowers laterally at the top. The flowers are about 6 inches long, white, and very fragrant.

952. Fritillaria thunbergii, *Miq.*, Jap. *Haru-yuri, Haha-yuri, Amigasa-yuri ;* a garden bulbous plant of the order Liliaceæ, growing to a height of about 1 ft., with long narrow leaves. The three leaves at the top are rolled up at the end. In spring it produces short peduncles from the axils of leaves, and droops one flower from each peduncle. The flowers are 6-petaled, bell-formed, light yellow with green veins, and purple spotted inside.

953. Fritillaria japonica, *Miq.*, Jap. *Koba-imo, Tengai-yuri ;* a small variety of the preceding, growing in shady places of valleys. It attains to a height of 3-4 inches. In spring it blooms on short peduncles produced from the axils of leaves, much resembling the former.

954. Fritillaria camtschatensis, *Gawl.*, Jap. *Kuro-yuri, Koku-yuri ;* it grows on high mountains of northern regions. It attains to a height of about 1 ft. In early summer it bears 6-petaled bell-shaped dark purple flowers facing laterally at the top of the plant. It is not a real lily. The bulbs are eaten by the native of *Yeso*.

955. Hemerocallis flava, *L.*, Jap. *Wasure-gusa ;* a perennial wild herb of the order Liliaceæ. In summer it grows to a height of about 2 fts., with several flowers at the top, blooming successively. The flowers are 6-petaled and reddish yellow. They open in the morning and fade in the evening. The flowers and young shoots are eaten as vegetables, being soft and sweet.

955. b. Hemerocallis flava, *L.*, fl. pleno., Jap. *Yaye-kwanzō, Yabu-kwanzō, Oni-kwanzō;* a double-petaled variety of the preceding, growing to a height of about 3 fts., with large long leaves and yellowish brown flowers. It is used in the same way as the preceding.

956. Hemerocallis minor, *Mill.*, Jap. *Beni-kwansō, Beni-suge;* a variety of *Wasure-gusa* (955), growing wild on mountains, and also much planted in gardens. In summer it grows to a height of about 1½ fts., and produces several yellowish dark-red flowers at the top.

957. Hemerocallis dumortieri, *Morr.*, Jap. *Hime-kwansō, Kisuge;* a small variety of *Wasure-gusa* (955), with golden yellow flowers in early summer.

957. b. Hemerocallis, Jap. *Zenteikwa, Setteikwa, Nik-ko-kisuge;* a species of Hemerocallis smaller than 955, but larger than the preceding. Its flowers are golden yellow.

957. c. Hemerocallis, Jap. *Yūsuge, Yoshino-kisuge;* a mountain variety of Hemerocallis flava (955), growing to a height of 3-4 fts. Its yellow flowers open in the evening, being fragrant at night, and fade in the next morning.

957. d. Hemerocallis flava, *L.*, var. foliis variegata, Jap. *Suji-kwansō;* it is princially planted in gardens. Its leaves are variegated with white longitudes.

958. Tricyrtis japonica, *Hook.*, Jap. *Hototogisu;* a perennial herb of the order Liliaceae growing in shady places of mountains. It attains to a height of about 1 ft. The leaves are generally spotted. From summer to autumn, it produces white 6-petaled flowers with purple spots. A variety with yellow flowers is called *Tamagawa-hototogisu*.

959. Funkia sieboldiana, *Hook.*, Jap. *Tōgibōshi;* a perennial garden herb of the order Liliaceae, producing many leaves from one root. In summer it grows to a height of 4-5 fts., and bears many white 6-petaled flowers in panicles at the top of the stem. The flowers do not open fully. A variety called *Tamanokanzashi* has narrow leaves, and its flowers open fully. Another variety called *Tokudama* is small, and its leaves are covered with white powder.

960. Funkia ovata, *Spreng.*, Jap. *Gibōshi;* a small wild variety of the preceding, growing to a height of 1-2 fts. In summer it bears purple or white flowers. There are many varieties. The petioles of all these varieties are edible.

960. b. Funkia japonica, *Spreng.*, var., Jap. *Sujigibōshi;* a garden variety of the preceding with yellow or white stripes on leaves.

961. Rhodea japonica, *Roth.*, Jap. *Omoto;* an evergreen herb of the order Liliaceae, growing in mountains of warm regions, but principally planted in gardens as a pot plant. The leaves are dark green, broad and about 1 ft. long. It shoots out a peduncle from the centre of several leaves to a height of 5-6 inches, and produces small flowers, being succeeded with a cluster of beautiful red berries. The leaves are various with different sizes, shapes, and variegations. They are admired on account of their beautiful evergreen leaves.

962. Plectogyne variegata, *Link.*, Jap. *Haran;* an evergreen garden herb of the order Liliaceae. The leaves are narrow or broad, and 2-3 fts. long. The narrow leaves stand straight, while the broad leaves inclined at the end. In spring it yields a dark purple flower near the roots, being succeeded with blue fruits as large as a finger's head.

963. Tofieldia nuda, *Max.*, Jap. *Hanajekishō, Iwazekishō;* a perennial herb of the order Liliaceae growing on rocks of valleys. The leaves are long and narrow. In summer its hoots a peduncle from the centre of the leaves and produces small white flowers in panicles.

964. **Scirpus lacustris,** *L.*, var. genuinus, *Gren.*, Jap. *Futoi, Tōi, Tsukumo, Marugama ;* a perennial herb of the order Cyperaceæ growing in ponds and marshes. The stalk is round, and grows to a heigt of 5-6 fts. In summer it produces several peduncles with many yellowish green flowers. The stalks are used to make mats called *Gama-mushiro*, and also used to make cords.

Lepironia mucronata, *Rich.*

965. ___ . Jap. *Inperai, Hirasuge, Nebikigusa ;* an evergreen herb of the order Cyperaceæ growing in the marshy places of warm provinces. The leaves are flat and round, being about 3 fts. long. In summer its stalk shoots forth from the centre of the leaves and blooms at the top. The leaves are used to make mats.

966. **Eriocaulon sexangulare,** *L.*, Jap. *Hoshi-kusa, Chikutōsō ;* an annual herb of the order Eriocaulonaceæ growing in paddy fields and other moist places. It shoots out many slender leaves from one root, and in summer it produces several peduncles which grow to a height of 2-3 inches, bearing white ball-flowers at the tip. The male and female flowers are found separately. A large variety has a height of about 1 ft.

967. **Typha japonica,** *Miq.*, Jap. *Gama, Kaba, Hiragama, Ohgama ;* a perennial herb of the order Araceæ growing in ponds and marshes. It resembles *Hime-gama* (339), but the leaves are broader, and the male and female flowers are attac̄d closely together with larger spadix. The use and quality are the same as the preceding.

968. **Acorus spurius,** *Schott.,* Jap. *Shōbu, Ayamegusa ;* a perennial herb of the order Araceæ growing in ponds and other moist places. The leaves are flat, narrow, and 4-5 fts. long, resembling swords. In summer it produces many small flowers in panicles from the axils of leaves. It is a custom of Japan to put

the leaves on the roof on the May-festival day. A kind of incence is made from the roots.

969. Acorus gramineus, *Ait.*, Jap. *Sekishō*; a small species of the preceding, growing in valley-rivulets. It is planted by water-sides to prevent sand from its sliding down. There are several varieties of planted Acorus. They are highly prized by amateurs as evergreen ornamental pot-plants. A few of them are described in the following articles.

970. Acorus gramineus, *Ait.*, var., Jap. *Arisugawa-sekishō*; a variety of the preceding with a beautiful dark green straight leaves.

971. Acorus gramineus, *Ait.*, var., Jap. *Birōdo-sekishō*; a dwarf variety of the preceding with tiny leaves less than 1 inch in length.

972. Miscanthus sinensis, *Anders.* Jap. *Susuki, Obana;* a perennial wild herb of the order Gramineæ. In autumn it shoots up leaves to a length of 5-6 fts. Panicles of flowers grow at the top, and the ripen seeds fly off by their own pappus. It is planted as one of the seven autumn herbs. The leaves are used to make rope.

973. Miscanthus sinensis, *Anders.* var. zebrina, Jap. *Takanoha-susuki;* a variety of the preceding. Its leaves have lateral white stripes like the wing of a hawk. A variety with longitudinal stripes is called *Shima-susuki*.

974. Miscanthus, Jap. *Ito-susuki;* a species of *Susuki* (972) with slender leaves about 2-3 fts. long.

975. Equisetum ramosissimum, *Desf.*, Jap. *Inudokusa, Kawara-tokusa;* a perennial herb of the order Equisetaceæ growing in sandy places near water, resembling E. hiemale (296), though smaller.

975. b. Equisetum hiemale, *L.*, Jap. *Tokusa;* this

herb (296) is evergreen growing in clusters, and planted among trees and rocks.

976. Gleichenia glauca, *Hook.,* Jap. *Urajiro, Shida, Oshida, Honaga;* an evergreen herb of the order Filices growing in warm regions. Its roots creep under the ground, and shoot up stalks at any place. The old stalk attains to a height of 4-5 fts. The leaves are light green on the upper surface, but white on the lower surface. It is used for the decoration of New-year's days.

976. b. Gleichenia dichotoma, *Willd.,* Jap. *Ko-shida, Ko-urajiro;* a small variety of the preceding, but the leaves differ slightly.

977. Cyathea spinulosa, *Wall.,* Jap. *Hego, Oni-hego;* an evergreen tree of the order Filices found in warm provinces. It grows to a height of 30-60 fts., and produces many leaves at the top. A leaf is about 7 fts. long, and the spores attach to its under side. The lower part of the trunk is covered with dark brown coarse hair, and it is used as a pot-plant by cutting it. The upper part of the trunk with the scars of fallen leaves is used for an ornamental timber.

978. Lomaria euphlebia, *Kunze.,* Jap. *Kiji-no-o-shida;* a perennial mountain herb of the order Filices. It produces several leaves from the old roots. The leaves are about 1½ fts. long, with small leaflets growing pinnately on both sides of a petiole, resembling a pheasant-tail. It produces leaf-like peduncles with small yellowish green spores underneath.

979. Davallia bullata, *Wall.,* Jap. *Shinobu;* a perennial fern with rhizomes creeping on rocks and decayed woods. The leaves are finely compounded, and about 7 inches long.

980. Adianthum pedatum, *L.,* Jap. *Kujaku-shida, Nuribashi, Kujakusō;* a perennial fern growing in valleys. On one petiole more than 10 leaflets shoot forth pinnately on both sides, resembling a peacock-tail. Its young leaves are scarlet, and its petioles are lustrous and dark purple. The petioles without leaflets are used to make brooms.

981. Adianthum monochlamys, *Eat.*, Jap. *Hakone-shida*, *Hakone-sō*, *Oranda-sō* ; an evergreen fern growing on steep rocks of deep mountain-valleys. The leaves are about 1 ft. long. The petioles and stalks are lustrous and purplish black. The leaflets resemble a duck-foot.

982. Pteris serrulata, *L.*, Jap. *Inomotosō*, *Tori-no-ashi*, *Keisokusō* ; an evergreen fern growing in shady places. The leaves are about 1 ft. long, growing in tufts. In summer the spores grow underneath the leaf-margin. A large variety is called *Ō-inomotosō*.

983. Pteris cretica, *L.*, var. alba-lineata, Jap. *Matsu-zaka-shida*, *Okina-shida* ; a variety of the preceding with white stripes in the centre of the leaves.

984. Asplenium nidus, *L.*, Jap. *Taniwatari*, *Ō-tani-watari* ; an evergreen fern, growing in shady places of warm regions. The leaves are broad and large without segments. Its large leaves are about 3 fts. long.

985. Scolopendrium vulgare, *Sm.*, Jap. *Ko-tani-watari*, *Taka-no-ha* ; an evergreen fern growing in mountains. It resembles the preceding, but small. The leaves are provided with petioles about 1½ fts. long, bearing the spores as in the preceding.

986. Lomaria nipponica, *Kunze*, Jap. *Shishigashira*, *Kusa-sotetsu* ; an evergreen fern growing in valleys with many leaves in cluster, expanding horizontally. The leaves grow pinnately on both sides of the petiole like comb-teeth, and are about 1 ft. long, generally coiling at the tips. The spore-bearing fronds grow separately.

987. Camptosorus sibiricus, *Rupr.*, Jap. *Kumanosu-shida* ; an evergreen fern growing on mountain-rocks. The single leaves are 6-7 inches long, and their tips grow slender, reaching to the ground and shooting roots. The spores grow on the back of the leaves.

988. Gymnogramme elliptica, *Baker*, Jap. *Iwa-hitode, Nikkō-shida;* an evergreen fern creeping on rocks and trees of mountains. The leaves grow pinnately on both sides of the petioles in the form of expanded fingers. The spores are attached to the back of the leaves.

989. Aspidum lepidocaulon, *Hook*, Jap. *Oridsuru-shida, Tsuru-sotetsu;* an evergreen fern growing in mountains of warm regions, with several leaves in cluster. The leaves grow pinnately on both sides of the petioles. The tips of the petioles grow slender, and reach to the ground to take roots.

990. Aspidum falcatum, *Sw.*, Jap. *Oni-shida, Ushigomi-shida;* an evergreen fern found on sea-coasts of southern provinces, producing several leaves from a root, with leaflets on both sides of the petioles. The leaves are lustrous and dark green, being about 1 ft. long. It grows on the rocks of *Ushigomi* in *Tokio*, whence the Japanese name is derived.

991. Polypodium buergerianum, *Miq.*, Jap. *Yanone-shida;* a perennial fern creeping on the rocks of mountains. Its leaf rembles the head of an arrow, whence the Japanese name is derived. The spores are spotted on the back of the leaves.

992. Polypodium lingua, *Sw.*, Jap. *Hitotsuba;* an evergreen fern creeping on rocks and decayed woods in warm regions. The leaves are narrow, and their backs are yellowish brown. They are about 1 ft. long.

993. Polypodium hastatum, *Th.*, Jap. *Uraboshi, Hoshihitotsuba;* an evergreen fern creeping on rocks and trees of mountains. The leaf is not dissected, but sometimes forms a trifid or difid. The spores grow on the back of the leaves, being arranged separately like stars.

994. Polypodium ensatum, *Th.*, Jap. *Kuriharan ;* a large variety allied to the preceding, growing on mountain-rocks. The leaves are about 1 ft. long. The spores grow on both sides of the main vein underneath the leaves.

995. Polypodium tricuspe, *Swartz*, Jap. *Iwa-omodaka ;* a variety of Polypodium lingua (992), with ternate leaves, growing on decayed woods of mountains.

996. Polypodium lineare, *Th.*, Jap. *Noki-shinobu, Yatsumeran ;* an evergreen fern growing on trees, rocks and roofs. The leaves are narrow and 4-5 inches long. The spores adhere on both sides of the vein underneath the leaves.

997. Nephrolepis tuberosa, *Presl.*, Jap. *Tama-shida ;* an evergreen fern growing in shady places of mountains in warm regions. It shoots many fronds from one tuft to a height of about 2 fts. The leaves grow pinnately on both sides of the petioles. The roots are slender, strong and straight, creeping over and sometimes under the ground. The underground roots are provided with many bulbs, from which the plants propagate.

998. Lygodium japonicum, *Sw.*, Jap. *Tsuru-shinobu, Samisen-dsuru, Kani-kusa ;* a wild perennial scandent fern. The stalk is slender like fine wire, being several feet long, with compound leaves. The leaves are finely dissected, and have spores on the back. The ripe spores look like brown sand, being used as a medicine called *Kaikinsha*. This plant is a curious species of ferns.

999. Ophioglossum vulgatum, *L.*, Jap. *Hanayasuri ;* a wild perennial herb of the order Filices. The rhizomes extend under the ground in every direction, and shoot forth stalks which attain to a height of 6-6 fts., with a spoon like leaf on each stalk. File-like spikes of spores are attached to the ends of the branches. The ripe spores produce very fine powder as in the preceding.

1000. Botrychium lunarium, *Sw.*, Jap. *Akinohanawarabi ;* a perennial ophioglossaceous fern growing in shady places of mountains. Its stalk grows to a height of 6-7 inches, being provided with a pinnate leaf. Small spores are attached to the ends of branches in panicles. This is the same species with B. ternaturn (91. b), and is rarely found.

1001. Lycopodium aloifolium, *L.*, Jap. *Nankaku-ran, Iwamomi ;* an evergreen herb of the order Lycopodiaceæ, growing on decayed woods of mountains in warm regions, drooping to a length of 8-9 inches. The stem is closely imbricated with small leaf-like scales. The spores are produced at the head of branches and the axils of leaves.

1002. Lycopodium sieboldi, *Miq.*, Jap. *Himoran, Iwahimo, Itofuran ;* an evergreen Lycopod growing on decayed trees of mountains in warm regions. Several drooping slender stems are about 1 ft. long, and divided into many branches. The small leaves grow closely together, and look like a cord. The organs of fructification are closely allied to the preceding.

1003. Lycopodium clavatum, *L.*, Jap. *Hikage-no-kadsura ;* an evergreen mountain Lycopod. Its vine creeps over the ground and takes roots everywhere. Some vine is about 10 fts. long, and is divided into several branches. Stems and branches are covered with small scale-like leaves. Its branches bear peduncles divided into 2 or 3, and produce spores which when ripe yield fine yellowish white powder. This powder is used to smoothen globes and boots.

1004. Lycopodium japonicum, *Th.*, Jap. *Mannen-gusa, Mannen-sugi ;* an evergreen Lycopod resembling the preceding, with standing stems divided into many branches, which produce spikes of spores at the top. The roots creep under the ground, shooting up the stems everywhere. This plant does not change its aspect, though it dries up.

1005. Lycopodium complanatum, *L.*, var., chamæ-cyparissus, *Al.*, Jap. *Asuhi-kadzura, Tsuru-hiba ;* an evergreen Lycopod growing on high mountains. It resembles L. clavatum

·(1003), but the leaves are small, wrinkled, and attached more closely.

1006. Selaginella caulescens, *Spring,* Jap. *Kata-hiba, Hime-hiba;* a perennial herb of the order Lycopodiaceæ growing on trees and rocks of mountains. Its roots creep and produce stalks everywhere, dividing many branches in layers. The stalk stands obliquely to a height of 6-7 inches. This plant is covered with small leaves closely put together like scales, producing spikes of spores amidst the leaves.

1007. Selaginella involvens, *Spring.* Jap. *Iwa-hiba, Iwa-matsu;* an evergreen selaginella growing on rocks of mounains. Several plants grow together, and produce many horizontal branches, which are covered with numerous scale-like leaves. The branches lengthen when it rains and shrivel when they face the sun. In autumn it produces fine spikes. There are many varieties planted in pots.

1008. Psilotum triquetrum, *Sw.,* Jap. *Matsuba-ran, Chiku-ran;* an evergreen herb of the order Lycopodiaceæ growing on rocks in shady places of mountains in warm regions. The roots creep in ground, and shoot stalks everywhere. It has a height of about 1 ft., and is divided into many branches. The leaves are thin and steril. It produces yellow small spores on the branches. There are more than 100 varieties.

1009. Polyporus niponicus, Jap. *Keishi, Saiwai-take, Mannen-take;* a fungus growing on decayed roots with cloud-like variegations on the pileus attached to the stipe. In summer it grows in the form of a Japanese writing-brush and gradually expands forming the pileus. The stem and upper part of the pileus are red, purple, or yellow, with a lacquer-like lustre, and the under part of the pileus is brown and coarse. It is hard like a cork and can be preserved for a long time. It is precious as a pot-plant. Sometimes it has two lagers of the pileus or it is divided into several branches.

1010. Polyporus, Jap. *Rokkakushi;* a variety of the preceding growing rarely in mountains. The stem in divided into many branches, but not provided with pileus, though it has a brown part at the head. The whole shape resembles antlers.

Note.—Ornamental plants are very numerous, and those described in this chapter are only a part. Among wild plants, there are many ornamental plants with beautiful flowers, variegated leaves, dwarf forms, etc. Many plants growing in mountains and valleys are also ornamental. Especially ornamental garden-plants are very numerous with leaves and flowers of various colours stripes, spots, variegations, sizes, shapes, etc., and some of them have more than hundred varieties; so among their varieties only one or two were described in this chapter. Ornamental garden-plants described in other chapters are Hibiscus manihot (357), Valeriana officinalis (436), Aconitum chinensis (482), etc.; ornamental pot-plants are Petasites japonicus (67), Caspicum longum (167), Orithia exypetala (258), etc.; and ornamental plants for vase-flowers are Brassica chinensis (48), Chrysanthemum coronarium (62), Pueraria thunbergiana (251), etc.

CHAPTER XXV.—Ornamental Plants for Covering and Shading the Ground.

This chapter contains the plants used for lawns, and those planted on river-banks, sea-coasts, mountain-cliffs, or mounds to prevent sand from falling down are also mentioned here.

1011. Ophiopogon japonicus, *Gawl.,* Jap. *Riūno-hige, Jano-hige;* an evergreen herb of the order Liliaceæ, being a veriety of O. spicatus (467) with narrow leaves. It grows wild in bamboo-woods and forests. The leaves are about a foot long, growing in tufts. In summer it produces peduncles, and bears 6-petaled purplish flowers in panicles, being succeeded with blue round pea-sized berries. It thrives well in shady places under trees.

1012. **Zoysia pungens,** *Willd.*, Jap. *Chōsen-shiba* *Yaye-shiba*, *Kōrai-shiba;* a perennial grass with very fine small leaves, creeping over the ground and taking roots at the joints. The leaves are 1-2 inches long, and amidst them produces small panicles of flowers, being succeeded with fine seeds. It is the best plant for lawn. A variety with tiny leaves is called *Chirimen-shiba*, which fades in cold weather.

1012. b. **Zoysia macrostachya,** *Fr. et Sav.*, Jap. *Shiba*, *No-shiba;* a large species of the preceding growing everywhere. It is planted on mounds, mountain-cliffs, etc., to protect earth from falling down.

1013. **Miscanthus japonicus,** *Benth.*, var., Jap. *Toki-wa-susuki*, *Kan-susuki;* an evergreen grass growing in bushes by sea-coasts of warm regions. The leaves are about 5-6 fts. long. In cold regions the leaves wither more or less, but most of them remain evergreen. In autumn it produces panicles of flowers. This plant is planted on sea-coasts to protect the sand from being washed away by waves. The sheath of the young leaves are made into ropes, and the young panicles are made into brooms.

1014. **Miscanthus sacchariflorus,** *Hack.* Jap. *Ogi*, *Ogi-yoshi*, *Umi-yoshi;* a perennial grass growing by water-sides and in plains. Its rhizomes creep under the ground, and produce stalks from each joint to a height of 5-6 fts. The leaves resemble those of Miscanthus japonicus (972), but have no sharp dissection on the edges. The panicles also resemble those of 972, but larger and longer. This grass is fitted to protect mounds from falling off.

1015. **Phragmites roxburghii,** *Kunth*, Jap. *Yoshi*, *Ashi;* a perennial grass growing in marshy places. Its roots creep under the ground and shoots up stalks to a height of about 6 fts., bearing panicles at the tops. The stalks resemble small bamboos, being slender, light and lustrous, and they are used to make blinds. The thickness of the stalk depends on the fertility of the ground. Those growing by sea-coasts are slender, flexible and strong. This grass is planted in water-sides to protect mud from being washed away by waves. Its young sprouts are edible. Those produced in *Udono*-village of Province *Setsu* are called *Udono-yoshi*, and are very famous for their large and long stalks. They are used to make *Shichiriki*, a musical instrument.

Note.—Those mentioned in this chapter are only a part. There are many other plants used for these purposes in many provinces. They may be divided into the following sections.

For gardens:—Zoysia pungens, *Willd.*, Ophiopon japonicus, *Gawl.*, Mazus rugosus, *Lour.*, var. macranthus, *Fr.* et *Sav.*, Ellisiophyllum reptans, *Max.*, Dichondra reptans, *Forst.*, mosses, etc.

For mounds:—Zoysia macrostachya, *Fr.* et *Sav.*, Miscanthus japonicus, *Benth.*, Pennisetum japonicum, *Trin.*, Carex morrowii, *Boott.*, Phyllostachys kumasasa, *Munro.*, *No-zasa* (Bambusa sp.), *Hōrai-chiku* (Bambusa sp.), etc.

For mountain-cliffs:—Rosa multiflora, *Th.*, Carex puberula, *Boott.*, Anthistiria arguens, *Willd.*, var. japonica, *Anders.*, Sporobolus elongatus, *R. Br.*, Cynodon dactylon, *Pers.*, several other Carex and Gramineae, etc.

For river-banks:—Phragmites roxburghii, *Kunth.*, Miscanthus, Cladium mariscus, *R. Br.*, Scirpus lacustris, *L.*, var. genuinus, Cyperus nutans, *Vahl.*, Acorns, Juncus, etc.

For sea-coasts:—Carex macrocephala, *Willd.*, Cyperus rotundus, *L.*, Miscanthus, Carex pieroti, *Miq.*, Ischaemum anthephoroides, *Miq.*, Hemarthria compressa, *R. Br.*, Ischaemum sieboldi, *Miq.*, Elymus arenarius, *L.*, Rosa rugosa, *Th.*, Vitex trifolia, *L.*, var. unfoliolata, *Schauer*, Rhaphiolepis japonica, *S.* et *Z.*, Juniferus littoralis, *Max.*, etc.

THE END.

明治二十八年九月七日印刷
同　　年九月十一日發行

發行所　東京市赤阪區赤阪溜池町一番地
大日本農會

編輯兼發行者　大日本農會幹事
赤阪區赤阪丹後町一番地
森　要太郎

印刷者　京橋區銀座四丁目一番地
島田　用定

印刷所　京橋區銀座四丁目一番地
八尾商店活版部
京橋區銀座四丁目一番地

www.ingramcontent.com/pod-product-compliance
Lightning Source LLC
Chambersburg PA
CBHW021839230426
43669CB00008B/1016